THE GREAT AIRSHIPS

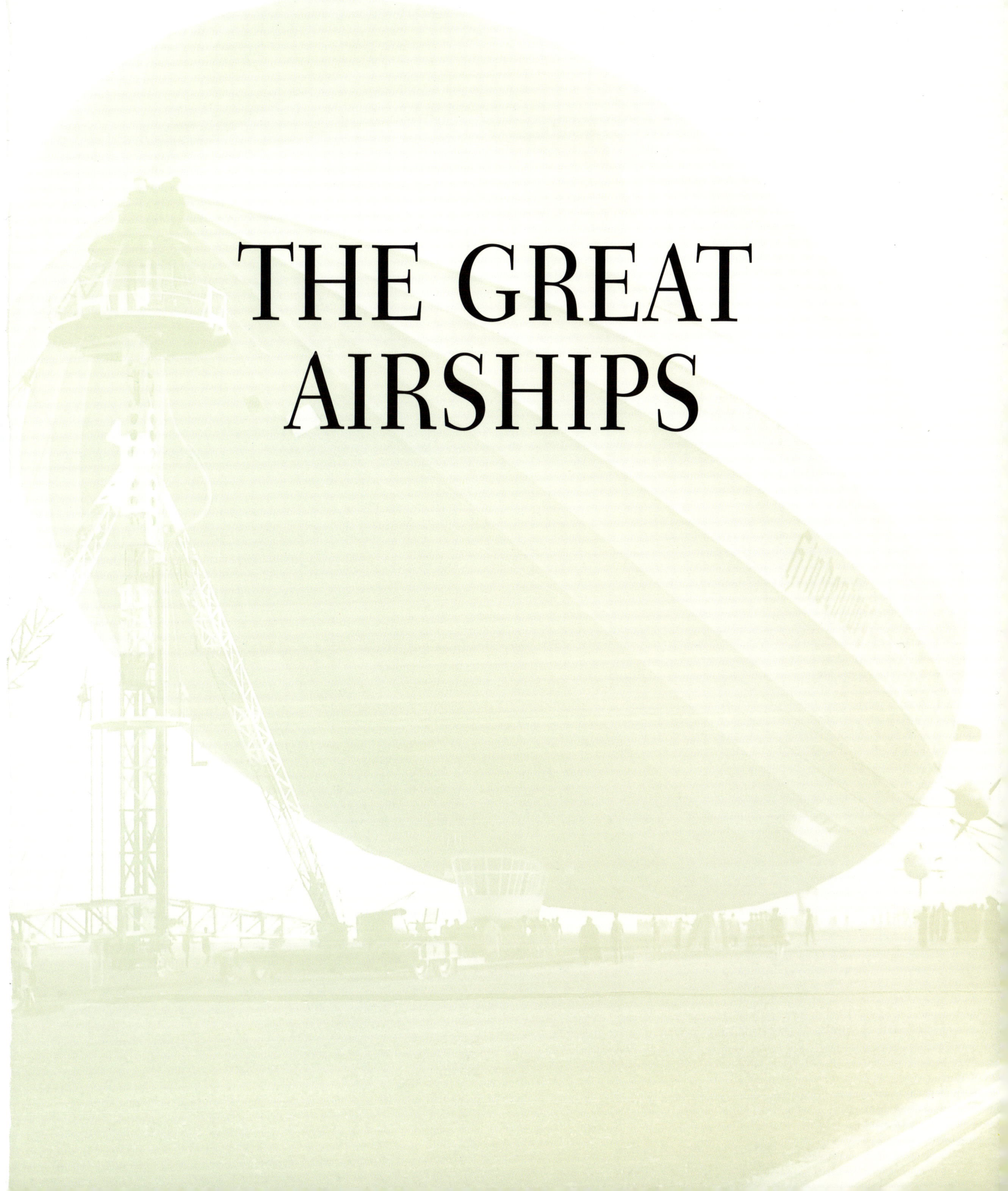

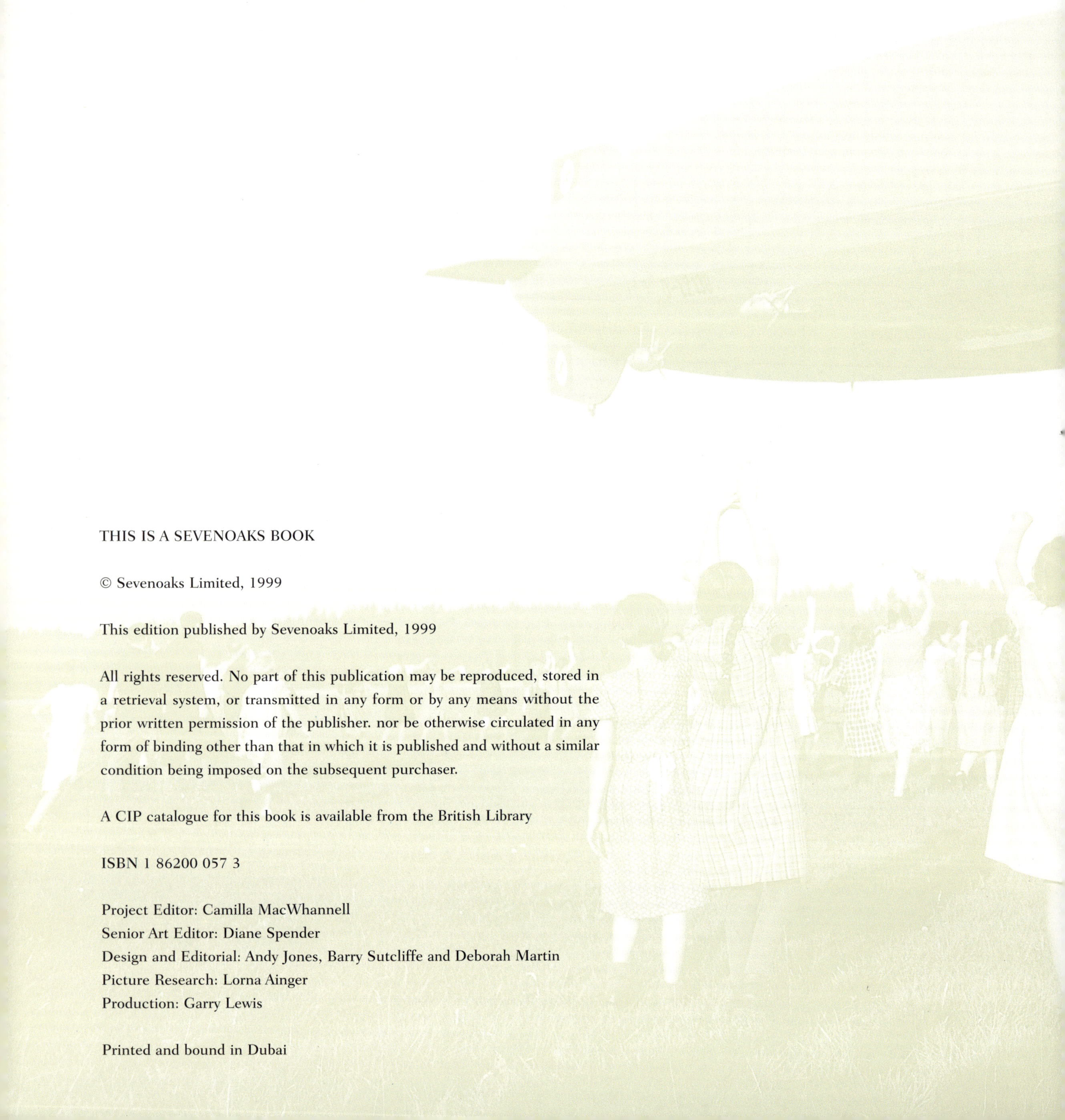

THIS IS A SEVENOAKS BOOK

This edition published by Sevenoaks Limited, 1999

A CIP catalogue for this book is available from the British Library

ISBN 1 86200 057 3

Project Editor: Camilla MacWhannell
Senior Art Editor: Diane Spender
Design and Editorial: Andy Jones, Barry Sutcliffe and Deborah Martin
Picture Research: Lorna Ainger
Production: Garry Lewis

Printed and bound in Dubai

THE GREAT AIRSHIPS

The tragedies and triumphs: from the Hindenburg to the cargo carriers of the new millennium

Mike Flynn

SevenOaks

Contents

Prologue

THE END OF A DREAM: THE *HINDENBURG* BURSTS INTO FLAMES BESIDE THE MOORING MAST AT LAKEHURST.

As a youthful announcer with the Chicago-based radio station WLS, Herbert "Herb" Morrison was used to covering run-of-the-mill events for his loyal band of listeners.

Community events, grand openings and the occasional visits of the great and the good were standard fare for Morrison and his sound engineer, Charlie Nehlson.

They had arrived at Lakehurst Field in plenty of time for the scheduled arrival of the great German airship the *Hindenburg* and chose to set up camp in a small aircraft hangar next to the enormous Hangar Number One, which had been built especially to hold airships. Herb checked his view from the hangar's picture windows while Charlie set up the recording equipment. In the 1930s, as-it-happens news items were usually recorded on to huge metal discs, which would then be taken back to the radio station and played on air for the benefit of the listening public. In those days, radio stations would compete to bring the news to their listeners as fast as they possibly could.

After a full day's frustrating wait, the *Hindenburg* finally made its approach for landing. "Here it comes, ladies and gentlemen," Morrison began brightly, "and what a sight it is. A thrilling one, just a marvellous sight. It is coming down out of the sky, pointed toward us and toward the mooring mast … The sun is striking the windows of the observation deck on the eastward side and sparkling like glittering jewels against a background of black velvet." And so he continued, relaying an almost routine account of what was happening before him, filling in the gaps between the action with flowery descriptions of the ship itself until the moment that every young reporter prays for – the big story.

THE GREAT AIRSHIP MOORED AT LAKEHURST FIELD, NEW JERSEY, ON ONE OF HER VISITS IN 1936.

Before his very eyes, Morrison saw emerge the biggest story of his entire life and must surely have wished that he had prayed for something else. In an instant, his calm, professional manner changed.

"It's burst into flames. Get this Charlie, get this, Charlie … Get out of the way, please, oh, my, this is terrible, oh, my, get out of the way, please! It is burning, bursting into flames and is falling on the mooring mast and all the folks we … this is one of the worst catastrophes in the world! Oh, it's four or five hundred feet into the sky, it's a terrific crash, ladies and gentlemen. Oh, the humanity…."

Thirty-five people lost their lives in the time it took Morrison to speak those words. The next day the world awoke to the first on-the-scene, press-covered air disaster in history and movie theatres were soon showing the shocking footage of the *Hindenburg*'s fall from grace. All who heard Morrison's broadcast would remember it for a long time, thankful that they had not been on the airship when it met its terrible end.

But something more than precious human lives were lost that day. With the death of the *Hindenburg* came the death of the great airship. America, Britain, France and Italy had all seen their dreams of conquering the skies in these magnificent craft turn into nightmares. Now Germany, the sole remaining builder of these huge, imposing structures, had also had its hopes dashed for ever. This was truly a sad end to what had started as the most beautiful and innocent of dreams less than 150 years earlier.

Chapter 1

Dawn Breaks

PHILIPPUS PARACELSUS, SWISS PHYSICIAN AND ALCHEMIST, AND ONE OF THE GREAT THINKERS OF THE RENAISSANCE.

The eighteenth century was a time of dreams. Since descending from the trees, the human race had come a long, long way. The proudest period in human history to date, the Renaissance, had seen an unprecedented flowering of the arts.

Magnificent works were produced during this period, the significance of which transcended even the noble qualities of beauty and grace. These were not the works of beasts of the fields, but the masterly achievements of a people who were at last learning to understand the world in which they lived, an understanding which had found its greatest expression in the works of artists such as Michelangelo. Here was a people who had learned to interpret their world and risen above the drudgery of a day-to-day struggle for existence.

Now the world was on the cusp of a new Renaissance, one which would see the seeds of science, first sown in the inquisitive minds of men such as Paracelsus and Pascal, blossom into the finest flowering of the human intellect. By the time this scientific and technological revolution was over, the human race would be master of all it surveyed. No longer would it dwell in ignorance, seeing the hand of God in all things. No longer would it be constrained to walk the Earth at a pace dictated by its own feeble frame. No longer would it cast an envious eye at the birds and wish that it too could take to the skies and join them. Before the century was over, the human race had taken its first steps on the journey to conquering the heavens, a trip that would take it, in just 200 years, beyond the Earth's atmosphere and on into the cold, silent vacuum of space.

As with so many of the great achievements of human civilization, the origins of human flight lie in that vast, unknowable country that is China. By the end of the first millennium, the Chinese had mastered the art of kite flying to such an extent that it had become a bizarre form of punishment. Persistent offenders were strapped to enormous kites and sent aloft on windy days as punishment for their crimes. Little could these early aviators have realized the contribution they were making to the science of human flight. (It would also have been quite beyond their comprehension to learn that one day people would choose to take to the air of their own free will strapped to similar devices, called hang gliders.)

While Leonardo da Vinci, the greatest of the Renaissance artists, dreamed of flying in all manner of improbable devices, including a helicopter of sorts, and an aeroplane equipped with impractical flapping wings, he kept his thoughts mostly to himself. In fact, his designs were not to surface until well after far more airworthy versions of his dreams had been conceived, designed and built by others. But what is clear from the great man's notebooks was that the idea of human flight was already gripping the minds of some of the finest intellects on the planet. Unfortunately, for Leonardo and his contemporaries, it was to be many years before these dreams came to fruition.

LEONARDO DA VINCI: AN ENGRAVING BASED ON A SELF-PORTRAIT.

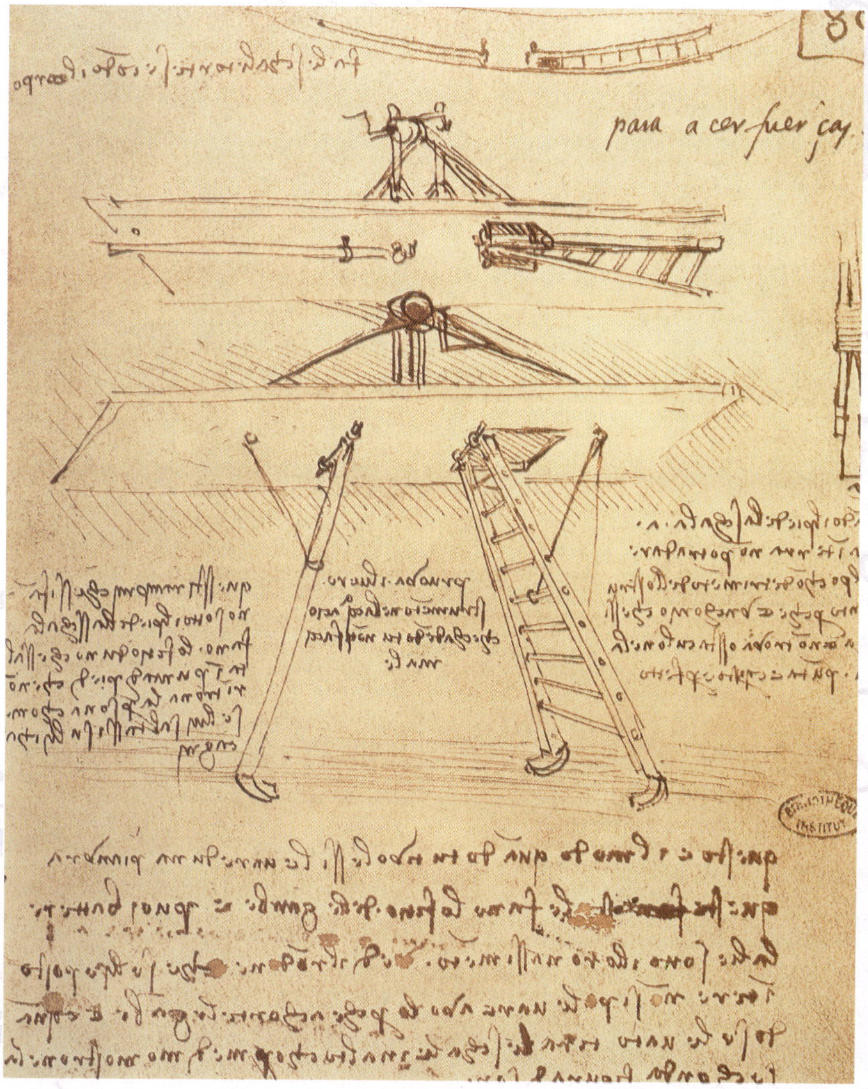

ONE OF LEONARDO'S MANY DESIGNS FOR A FLYING MACHINE, INCLUDING LADDERS FOR DESCENDING.

But here, before people's very eyes, the Montgolfier brothers and their fire had breathed life into the bag, or balloon.

THE BALLOON THAT CARRIED DE ROZIER AND THE MARQUIS D'ARLANDES IN THE FIRST MANNED FLIGHT, SHOWN HERE BEFORE LIFT-OFF.

Brothers Joseph-Michel and Jacques-Etienne Montgolfier were unlikely pioneers. Their parents owned a very successful paper-making business, but still found time away from the chores of commerce to produce 16 children, two of whom were to earn well-deserved places in the history books. No one knows what sparked the brothers' interest in aviation; they certainly had no scientific training and knew little of the mysteries of engineering. We do know, however, that they would have had neither the time nor the resources to pursue their hobby had it not been for the income they received from their parents' factories.

On a sun-scorched June 4, 1783, with France about to enter the bloodiest period of its turbulent recent history by means of revolution, a crowd gathered in the market square in the small town of Annonay to watch the Montgolfier brothers demonstrate their new invention. No one knows whether or not the crowd came to mock the brothers' efforts or if they were merely a loose collection of locals who simply happened to be going about their business when they stumbled upon a little local diversion, but we can be sure that for everyone present in the square it would be a day they would remember for the rest of their lives.

Before the bemused onlookers, the brothers produced what appeared to be a large and rather fetchingly decorated bag. They proceeded to set fire to a bundle of straw and held the bag over the resulting bonfire.

In an age when anything that the human imagination can conceive of is given life on cinema screens and televisions across the world, it is hard to imagine the effect upon the crowd of what happened next. In the eighteenth century, anything that flew did so because God willed it to do so. But here, before people's very eyes, the Montgolfier brothers and their fire had breathed life into the bag, lifting it into the air and sending it on its journey to the heavens. That day in June marked the first real step on the quest to conquer the skies. Yet what had seemed so magical and mysterious was really just the first simple public demonstration of the principles of lighter-than-air flight.

The Earth is shrouded in a thick, gaseous atmosphere that, for the most part, becomes progressively thinner the further one gets from the surface of the planet. When the brothers filled their bag – or balloon as everyone was calling it – with hot air, they were actually filling it with a gas that was less dense than the surrounding atmosphere. (Because the air in the balloon had been expanded by the heat from the burning straw, less of it was required to fill the balloon.) Just as an air bubble will rise through water, so the balloon, now filled with hot air, rose through the atmosphere until it reached a point where the density of the air it contained matched that of the surrounding atmosphere – in this case at around 3,000 feet.

The science of lighter-than-air flight mattered little, however, to the crowds who flocked to see the Montgolfier brothers demonstrate their new invention at markets and fairs all over France. What they were really interested in was the whole new world of possibilities that the demonstration presented, not least for human flight. This last point was certainly not missed by a couple of wealthy young men who saw the brothers demonstrate their invention in Paris the following month.

The Marquis d'Arlandes, a French aristocrat with plenty of time on

Pilâtre de Rozier who, with the Marquis d'Arlandes, seized on the idea of the Montgolfier brothers and became the first human balloonists.

his hands, had watched the Montgolfier brothers send three farmyard animals aloft in a hot-air balloon. With a grin, he had turned to his good friend Pilâtre de Rozier and both men had known at once what the other was thinking. With uncharacteristic diligence, the pair set about becoming the first humans to take to the skies in the new invention.

And so it was that on November 21, 1783, the men left the sharp, hard frost on the ground and floated up into the air. Once there they were captured by the wind and whisked off on a five-and-a-half mile adventure over the beautiful city of Paris. That the whole notion of human flight was beyond the comprehension of most people was demonstrated perfectly when they eventually came back down to earth, in a field on the outskirts of the city. The sight of the men drove a group of peasant farm labourers to fall to their knees in fear and wonder, thanking God for his kindness in sending two such remarkably well-dressed angels to watch over them.

Left: A 'Montgolfière', one of the hot-air balloons invented by the Montgolfier brothers, causes crowds to stare in amazement.

The year 1783 was clearly a momentous one in the history of flight. No sooner had the Montgolfier brothers demonstrated the principle of lighter-than-air flight, and d'Arlandes and de Rozier taken to the skies in a hot-air balloon, when another Frenchman, this time the eminent physicist Jacques-Alexandre Charles, decided to test the idea that lighter-than-air flight could be achieved using a gas other than heated air.

In principle, any balloon filled with a gas that is less dense than the surrounding air should have a good chance of getting off the ground. Charles reasoned that rather than going to all the bother of filling the balloon with artificially heated air, it would be more practical to fill the balloon with a gas that is naturally less dense than the surrounding air, even at normal temperatures.

The first hydrogen balloon, invented by Jacques-Alexandre Charles, took to the skies over Paris in 1783.

In this day and age, Charles would have tested his idea using a computer simulation. But eighteenth-century science afforded no such luxuries. With no health and safety executive to mollify, or commercial sponsors to impress, Charles merely attached a large basket to an even larger balloon, filled the balloon with hydrogen – the lightest gas of all – helped his good friend and colleague Nicolas Robert aboard and cut the rope that had been tethering the balloon and basket to the ground.

History does not record the words that must have passed between the two men at this point. This is probably just as well, for not only did Charles' theory work in practice, it worked rather better than expected. Before the men could comprehend what was happening, their craft had shot up over a mile into the sky. It was at that moment, on a cold December day, that true lighter-than-air flight came into the world kicking and screaming – along with its inventor.

One can only wonder at the courage of a man such as Jacques-Alexandre Charles. Not only did he continue to carry out research into lighter-than-air flight – once he had got over the shock of his first attempt – he also remained a ballooning enthusiast well into his old age.

Unfortunately, the progress of lighter-than-air flight had reached something of an impasse. Although the skies over Europe were soon filled with daring young men and women keen to get their first taste of aviation, no means could be found to control these early balloons. Once in the air, the craft and its occupants were held captive at the whim of the winds, blown wherever fate (and the weather) might take them. While this might be an inconvenience for civilian fliers, it caused far more serious problems for military balloonists.

THE BALLOON *INTREPID* IS INFLATED BEFORE THE BATTLE OF FAIR OAKS DURING THE AMERICAN CIVIL WAR; SUCH BALLOONS WERE USED FOR RECONNAISSANCE PURPOSES.

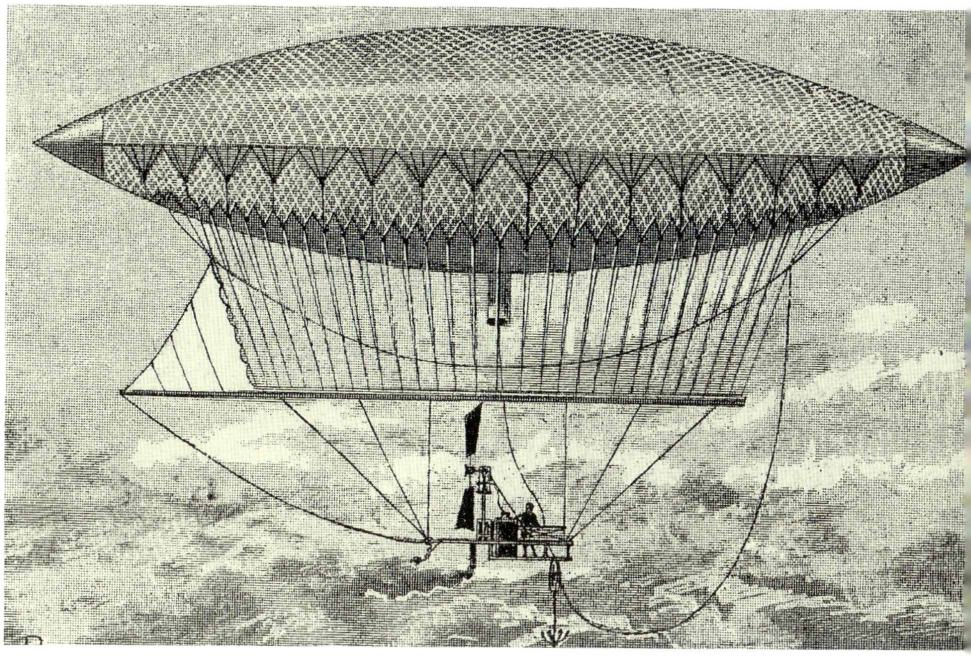

THE FIRST DIRIGIBLE BALLOON, INVENTED BY HENRI GIFFARD, WAS FITTED WITH A PROPELLER AND A STEAM ENGINE, ENABLING THE OCCUPANT TO CONTROL THE CRAFT'S DIRECTION OF FLIGHT.

As early as 1794, the French army had its own corps of balloonists, who went under the wonderfully romantic name of the Aérostiers. Their role was to act as reconnaissance for the French army, using their lofty position to gain a bird's-eye view of the battlefield. Unfortunately their activities were severely limited by the fact that the balloons had to be tethered to the ground, so that the Aérostiers could be recalled from on high to give their reports and to ensure that they did not drift off away from the field of battle or, worse still, into the enemy's hands. Until some means could be found of controlling the craft in flight, the military balloon stayed firmly tethered to the ground, where it remained for many years. Despite this, the reconnaissance balloon found a useful role with any number of armies over the next century or so, including both sides during the American Civil War. By then, of course, the race was truly on to find a way of producing a controllable, untethered craft that was lighter than air. Every military power, painfully aware of the advantages that such a craft would offer in battle, was putting its energies and huge financial resources into developing a steerable balloon.

Even before the terrible civil war had broken out in America, a Frenchman, Henri Giffard, had tested what he called his dirigible (from the French verb *diriger*, meaning "to steer"). Giffard's dirigible was a hydrogen-filled balloon that was shaped more like a cigar than the by now traditional "bag of air", but what really distinguished it from previous designs was the fact that it was fitted with a steam engine and a propeller. Basing the design on that of a ship, Giffard hoped to propel his "airship" through the skies as one might more normally drive a ship through water. His dirigible was, for the time, of enormous proportions. Because the steam engine that Giffard used to power his craft weighed in excess of 350 pounds, he was forced to use an enormous quantity of hydrogen to keep it in the air. The cigar-shaped bag that held all of this gas was, in the end, around 144 feet long.

In 1852, he took to the skies over Paris in the first public demonstration of his steerable lighter-than-air craft, managing a very

Showers of meteors observed by Giffard and Fonvielle during a flight in the balloon *L'Hirondelle*.

creditable 20-mile journey over the city, before leaving a crowd of open-mouthed onlookers in his wake as he sped away at the craft's top speed of six miles per hour.

To state that six miles an hour was not a terribly impressive speed is not to do Giffard a disservice. Henri Giffard's design was sound in every sense, but like other balloon designers of his day he was hamstrung by the available technology. Even though the steam engine that Giffard used for his craft was considered to be at the cutting edge of locomotive design, it none the less weighed a great deal. Its power output was the greatest available in an engine that size, but even so, a top speed of six miles per hour is simply not adequate for controlled flying in wind speeds of seven or eight miles per hour, or more. The age of the airship was going to have to wait until someone could come up with a better engine.

The wait appeared to be over when, in 1872, a German engineer by the name of Paul Haenlein took Giffard's dirigible design and added the recently invented, and considerably lighter, internal combustion engine. In a masterpiece of lateral thinking, he also managed to provide fuel for the engine by tapping into the hydrogen gas that filled the balloon, thereby greatly reducing the overall weight of the craft. Inspired as this idea was, however, it greatly reduced the flying time of his dirigible, which had a tendency to lose buoyancy as the hydrogen gas was gradually used up.

The marriage of electricity and magnetism, which had taken place earlier in the century, led to the fitting of an electric motor to the dirigible in 1883, when French brothers Albert and Gaston Tissandier attached just such a device to their craft. Their efforts were followed by those of

Alberto Santos-Dumont circles the Eiffel Tower in his airship *Model 6* in 1901. This won him the Deutsch Prize for the first flight from St Cloud to the Eiffel Tower and back in under 10 minutes.

Paris-based Brazilian Alberto Santos-Dumont, who took full advantage of the gasoline-powered internal combustion engine to set records over France's capital city with a dirigible of his own design.

Outstanding as these individual achievements were, however, they soon paled when placed next to the work of perhaps the greatest dirigible designer of them all – Count Ferdinand von Zeppelin.

Having begun his career as an officer in the Prussian cavalry, Zeppelin was driven to fight as a volunteer for the Union army during the American Civil War. He no doubt had many adventures during his time in America, but the key event in his personal experience of the war was his first encounter with a balloon, which was being used as a means of spying on the Confederate forces. Zeppelin was smitten. He became entirely fascinated by the craft, to the extent that he was to devote the remainder of his life to improving on the design that he had first encountered on a foreign battlefield. On his return to Germany he set about building the biggest and best dirigible that could be managed with the available technology.

Count Ferdinand von Zeppelin, probably the best known of all dirigible designers.

Zeppelin was quick to realize that his dirigible would have to be more rigid than those already in existence if he was to achieve the kinds of speed he had in mind for his craft. In what was his first and, some might argue, greatest contribution to the design of the dirigible, he added a very lightweight rigid frame to the elongated balloon shape he had adopted from Henri Giffard. This meant, in theory at least, that he would be able to control the craft more easily and that it was far less likely to buckle under stress. It would also allow him to build much larger dirigibles than had previously been possible without the addition of the lightweight frame. Zeppelin's design looked sound enough on the drawing board, but the time soon came to put his ideas to the test.

As the world entered the twentieth century, Zeppelin took his prototype dirigible to Lake Constance in Germany for its maiden flight. *Luftschiff Zeppelin - Ein*, or *LZ1*, took to the skies on July 2, 1900. Alas,

it had not been airborne for very long when disaster struck and the airship drifted slowly down to the surface of the lake. Zeppelin's disappointment was made all the worse when the *LZ1* collided with a safety buoy in the water and was ripped to shreds.

Undeterred, Zeppelin returned to the drawing board, where he remained for five years before launching the *LZ2*. He was not a man who could be easily beaten, especially as he was so firmly convinced that his design was viable, and he continued to refine his airship design. On July 4, 1908, as America, the country that had been the birthplace of his tremendous enthusiasm was in the midst of celebrating its independence, Count Ferdinand von Zeppelin set off on the maiden flight of his latest creation, the *LZ4*, an airship the like of which the world had never seen before.

Count Ferdinand von Zeppelin set off on the maiden flight of his latest creation, the LZ4.

At the time there was nothing to rival it in size. The *LZ4* required over half a million cubic feet of hydrogen gas to fill its vast balloon, which was a remarkable 446 feet long. On its maiden flight, the *LZ4* was flown for 12 hours over Switzerland at a continuous 40 miles per hour. The age of the airship had arrived in magnificent style.

Zeppelin's *LZ4*, a massive balloon measuring 446 feet in length, ready for its maiden flight in 1908.

Over 34,000 people took their first flight in one of Zeppelin's airships between 1910 and the outbreak of the First World War in 1914. This was a period that saw a huge increase in the number of airships being built, but this fevered activity was as nothing compared to the burst of airship building that took place during the war.

Between 1914 and the end of the war in 1918, Germany alone acquired 88 military airships. By now nearly all airships were referred to by the generic name "zeppelin", regardless of origin, which speaks volumes about Count Ferdinand von Zeppelin's contribution to the design of the craft. But although Zeppelin's motives in building his airship must have included the desire to see Germany become the most powerful military force in the world, it was in the post-war period that the airship could truly be said to have thrived. As the war in Europe ended, and people set to work rebuilding their shattered lives just as countries attempted to rebuild their economies, the new world of aviation was to become one of the most dynamic and attractive areas of activity.

Chapter 2

The Golden Age

ONE OF THE EARLY AEROPLANES BUILT BY THE WRIGHT BROTHERS IN FLIGHT AT KITTY HAWK, USA.

It is one of the unfortunate truths of any history of technology that war often proves to be the greatest spur to progress.

Wilbur Wright at the controls of his aircraft in 1908.

Everyone remembers that the Wright brothers built the first practical aeroplane, the *Wright Flyer*, but few seem to be aware that it flew for only 12 seconds on that first flight, or that the total flying time of the aeroplane during its entire lifetime was less than the time taken to serve a meal during a modern commercial flight. Serious development of the aeroplane did not begin in earnest until the Wright brothers took their invention to the American military, who agreed to fund their work just so long as the brothers could produce a reconnaissance aeroplane capable of "carrying a pilot and an observer for 125 miles at a speed of no less than 40 miles per hour". That the Wright brothers were able to provide just such an aircraft, barely more than a year after being commissioned to do so, says much about the influence of the military on the progress of the development of flying machines.

At the time the Wright brothers were demonstrating their invention, the airship was still viewed as the most likely weapon of any war in the skies. Here was a tried-and-tested technology which appeared to be the most probable way forward as the world began to prepare for the war which many knew, in their hearts, was almost inevitable. At the start of the First World War Germany had only 10 zeppelins, but added another 78 military airships during the course of the war. These were used mostly for reconnaissance work and for the aerial bombardment of assorted European cities, with London receiving the rather dubious honour of being the first capital city in the world to be attacked in this way. But, interestingly and uniquely, it was the post-war period that saw the greatest developments in airship technology.

Perhaps it was the sheer beauty of the airship, or the stately manner in which it made its way across the skies, that brought out the very best

THE SIGNING OF THE TREATY OF VERSAILLES IN THE HALL OF MIRRORS IN 1919; THE TERMS OF THE TREATY REQUIRED GERMANY TO SURRENDER HER AIRSHIPS TO THE ALLIES.

in the engineers who laboured on these galleons of the air. Whatever their motives, the post-war period saw some of the world's finest minds at work on producing ever bigger and better airships. These men were driven, or so it would seem, not so much by the desire to create more efficient killing machines, but to explore the boundless possibilities of human ingenuity. This was a period of records and races, of astounding achievements and remarkable advances. It was truly the golden age of the airship. This golden age did not, however, have a promising start.

Under the terms of the Treaty of Versailles, drawn up by the Allies at the end of the First World War, Germany was required to make reparation for the cost of the conflict. Each of the victorious countries was to receive, among other things, their share of Germany's stock of rigid airships or cash to a similar value. The whole idea of surrender had been unacceptable to many Germans, but to be humiliated further in this way was more than some could bear. Before America received its share of the airships – seven in total – they were wrecked beyond repair by an airborne unit of the German Navy. Just when it seemed that things could get no worse for Germany's struggling airship industry, the Allies refused to allow Germany to build any more military airships and placed an impossibly restricting limit on the size of any commercial airships that might be built. By 1920, the once mighty Zeppelin Company had become little more than a manufacturer of pots and pans.

With the death of Count Ferdinand von Zeppelin in 1917, the company had become caught up in a power struggle between rival factions, a conflict that was not resolved until 1924. In the meantime, both the British and the Americans were at work attempting to produce their own airships. These, it must be said, were little more than pale imitations of those produced by Germany during the war. One airship that did manage a creditable performance, however, was the British-built *R34*.

The possibility of an Atlantic crossing by airship had been considered by many people. By the end of the war distance already seemed to be no object. In November 1917, the German-built *L59* had flown to East Africa in an attempt to provide relief for the beleaguered German Army. It had been turned back at the last moment by a bogus radio message, sent by the British, informing the crew that the German Army in Africa had surrendered and that it would be dangerous and pointless for them to attempt to complete their mission. By the time the *L59* had returned to its base it had travelled continuously for 4,200 miles, covering the distance in just 95 hours and arriving back at base with enough fuel left for a further 64 hours of flight. By comparison, the trip across the Atlantic in the *R34* should have been a breeze. Unfortunately for all concerned, the breeze turned into a gale.

By the time the L59 had returned to its base it had travelled continuously for 4,200 miles.

A mishap shortly after take-off from Scotland had resulted in the *R34* venting hydrogen, which meant that the airship was essentially going to be overweight for the entire trip. In order to compensate, the officer in charge had increased lift by using the airship's elevators. Although this ploy worked, it had a dramatic impact on fuel efficiency. Added to this was the need to make numerous course changes in order to avoid thunderstorms. The *R34* finally reached Long Island in America after spending over 108 hours in the air – a record at the time – but with just enough fuel left for two hours of flight. The return journey, some of which had to be made using only three of the airship's four engines, benefited from a strong tail wind and saw the *R34* land in England after a little over 75 hours in the air.

The Americans, who were having trouble producing their own airship, were so impressed by the transatlantic flight of the *R34* that they commissioned the British to produce a new airship for the US Navy. The proposed airship was based partly on a German design known as a "height-climber", an airship of lightweight construction intended, as the name implies, for high-altitude flying rather than speed or manoeuvrability. This new airship, called the *R38*, took to the skies in 1921 and was the largest airship built so far, being just short of 700 feet long and with a hydrogen capacity approaching three million cubic feet. On its fourth test flight, with a group of American observers on board, the airship was put though a number of severe turns, the last of which caused the ship's lightweight structure to buckle and snap.

THE WRECKAGE OF THE *R38* IS RETRIEVED FROM THE RIVER HUMBER AFTER THE AIRSHIP'S FATAL TEST FLIGHT IN 1921.

The front end of the *R38* exploded and crashed into the River Humber below, killing all the men who were caught in that section. The rear of the airship dropped on to a sand bank in the river, killing all but four of the remaining crew and effectively destroying Britain's involvement in airship production for the next ten years. It seemed as though the disaster caused all concerned to lose heart, but just when it appeared that the days of the airship were numbered, Hugo Eckener, of the floundering Zeppelin Company, came up with a rescue plan.

Aware that America was still owed an airship under the terms of the Treaty of Versailles, and knowing that they were unlikely to receive one because of the German Navy's destruction of seven airships at the end of the war, he approached the American Military Commission in Berlin with an offer. He would see to it that the Zeppelin Company, now in civilian hands but once acknowledged as the world leader in cutting-edge military airships, would build them the biggest and best airship ever to take to the skies. With almost indecent haste, and against the wishes of the other Allies, the Americans jumped at the offer. The signing of the contract between the Zeppelin Company and the American Navy for the construction of an airship, the *LZ126*, to be delivered in 1923, was the first step back on the road to recovery for the company that had been founded by the great Count Ferdinand von Zeppelin.

The construction of the *LZ126* marked the revival of the Zeppelin Company after the First World War.

In the meantime, the Italians had been busy building up their own airship industry. In 1925, the Norwegian polar explorer Roald Amundsen approached leading Italian airship designer Umberto Nobile and asked him to pilot one of his own airships on an expedition across the Arctic from Spitsbergen to Alaska via the North Pole. (At this time the north polar region had resisted all attempts at exploration.) The Italian readily agreed to undertake the mission and on May 11, 1926, Amundsen, Nobile and his crew set off on their journey across the top of the world.

They reached the North Pole with relative ease, but the journey on to Alaska proved to be a perilous one. Freezing fog and ice combined with sheer exhaustion on the part of the crew led to Nobile's decision to land the airship in a small bay on the Alaskan coast. Despite the fact that they had failed to reach their original destination, Amundsen, Nobile and the crew could derive great satisfaction from the fact that their flight had been a success in every other respect.

Back in Germany, Hugo Eckener and the Zeppelin Company had delivered the airship that the American Navy had requested and, with a relaxing of the rules imposed by the Allies after the First World War, were back in the business of building airships. The 4,200-mile African trip of the *L59* in 1917 had convinced Eckener that a regular transatlantic flight by airship was a viable proposition. With this in mind, he set to work finding backers for just such a craft, which was given the serial number *LZ127*.

Even before he had raised all the money needed for the project, Eckener began work on constructing the new airship at the Zeppelin works in Friedrichshafen. On September 18, 1928, the *LZ127*, by now glorying in the new name of *Graf Zeppelin* in honour of the founder of the company, began the first of a series of test flights. These were completed without a hitch and so the airship on which the Zeppelin Company was pinning all of its future hopes was sent on its maiden voyage to New York with Hugo Eckener at the helm.

At 775 feet in length, and with a hydrogen capacity of 3,037,000 cubic feet, the *Graf Zeppelin* was the largest airship that had ever been built, and only just cleared the sides of the shed in which it was constructed. It was capable of carrying 20 passengers and even had a dining room and kitchen on board. Initially all went well with the flight, but on the morning of the second day, with the airship well away from land, a storm blew in.

THE NORWEGIAN EXPLORER ROALD AMUNDSEN IN THE CONTROL ROOM OF THE AIRSHIP *NORGE*, IN WHICH HE AND UMBERTO NOBILE FLEW OVER THE NORTH POLE.

THE ITALIAN AVIATOR AND ENGINEER UMBERTO NOBILE WAS A LEADING AIRSHIP DESIGNER IN THE 1920S.

The airship, suddenly dwarfed by the weather and the ocean below, was pitched and tossed across the skies like a leaf caught in a gale. The storm tore a huge hole in the fabric covering one of the tail fins. With weather conditions worsening, and the airship losing altitude, Eckener was forced to radio the US Navy for assistance. By now, despite the obvious dangers involved in such an operation, several members of the crew were in the exposed tail section of the airship, attempting to repair the damage. This was a grim task, carried out in the sure and certain knowledge that any of them could be blown to their death at any moment. Clinging on for dear life, the remarkable crew managed to repair most of the damage and eventually the *Graf Zeppelin* was able to carry on with its journey.

Before the rescue vessel could be recalled, however, news of the troubled airship had reached the American press. As no word was heard from the *Graf Zeppelin* until the rescue ship was cancelled, many Americans believed that it must have been lost at sea. News that it had survived its ordeal was cause for great rejoicing and also guaranteed a huge crowd for its arrival in New York. After nearly 112 hours in the air, and a journey of over 6,000 miles, the magnificent airship appeared over the city in the middle of the afternoon, bringing office workers and shopkeepers alike out on to the streets. The crew were treated to a ticker-tape parade through the city and at last it seemed that Count Ferdinand von Zeppelin's company was back in business.

HUGO ECKENER, IDENTIFIED BY A CROSS IN THE PICTURE, AND COMPATRIOTS ARE DWARFED BENEATH THE MASSIVE BULK OF THE *GRAF ZEPPELIN*.

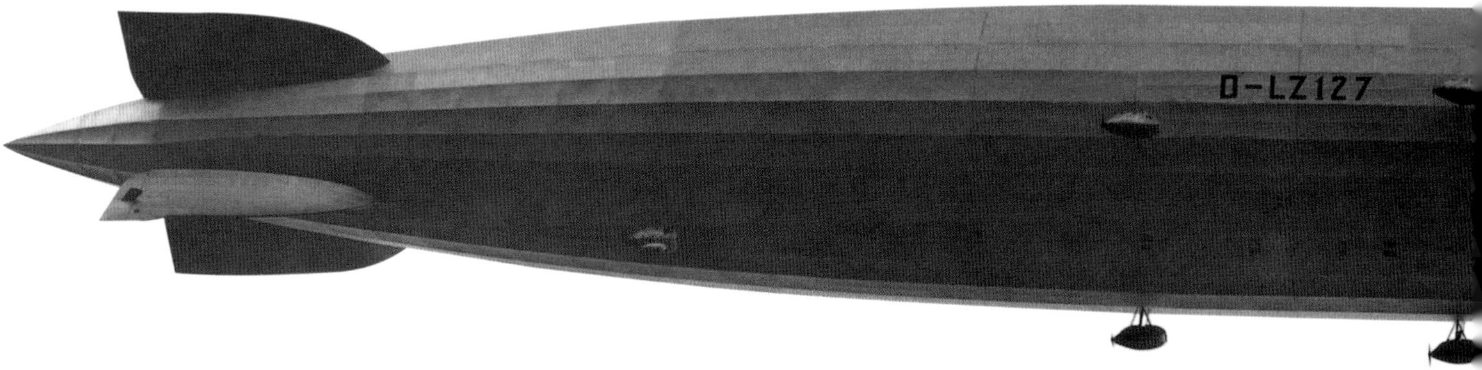

WITH A LENGTH OF 775 FEET, THE *GRAF ZEPPELIN* MADE AN IMPRESSIVE SIGHT IN THE SKIES.

The airship's dining room ressembled a high-class restaurant.

Despite having made a successful crossing of the Atlantic, Eckener was still experiencing difficulties in finding backers for his airships. With the German government reluctant to put much public money into his projects he was forced to look elsewhere. In a daring calculated gamble, Eckener decided to propose a new project for his magnificent airship – a non-stop trip around the world. Ingeniously, he raised much of the money needed for the project by offering "exclusive" rights to the story of the trip to a number of newspapers: William Randolph Hearst put up $100,000 of the estimated $250,000 needed in return for the worldwide English language rights to the story.

On August 15, 1929, the *Graf Zeppelin* lifted off from its base at

Work being carried out on the airship's frame at the Zeppelin works at Friedrichshafen.

WILLIAM RANDOLPH HEARST, ONE OF THE BACKERS OF THE GRAF ZEPPELIN'S ROUND-THE-WORLD FLIGHT IN 1929.

The Graf Zeppelin crossed the 5,000-mile expanse of the Pacific Ocean in less than three days.

Friedrichshafen and began its historic flight around the world. (It had earlier been to Lakehurst, New Jersey, at the insistence of William Randolph Hearst, who had been determined that the flight would start from America.) By August 19 the airship had reached Tokyo, where its arrival was greeted by an enormous crowd. After refuelling, the ship took to the skies again. With a typhoon behind it, the *Graf Zeppelin* crossed the 5,000-mile expanse of the Pacific Ocean in less than three days, arriving in San Francisco just as the sun was setting. From there it went on to Los Angeles, stopped briefly in New York and then flew home to Friedrichshafen.

The mighty *Graf Zeppelin*, with Hugo Eckener at the helm, had completed the round-the-world flight in exactly 21 days, 5 hours and 54 minutes. History had been made, the Zeppelin Company was restored, stronger than ever, to its position as the world's premier producer of airships, and any doubts about the future viability of airship travel were, it seemed, laid to rest once and for all. Lighter-than-air flight had come a long way since the Montgolfier brothers first sent their simple balloons skyward at the end of the eighteenth century. But the human cost of these achievements had been high, and as the world entered a new decade the price was set to climb higher still.

THE *GRAF ZEPPELIN* MOORED AT LAKEHURST, NEW JERSEY, BEFORE SETTING OFF ROUND THE WORLD.

Chapter 3

Icarus Falls

"The Fall of Icarus", attributed to Jacob Grimmer; for some, a symbol of the futility of human attempts at flight.

Wherever there is glory there is also pain. There are risks to be faced in every field of human endeavour and it is usually true that the more ambitious the endeavour, the greater the risks involved.

So much of our knowledge of the world has been gained at the expense of human life, as if nature has been playing jokers all along, offering a tantalizing glimpse of her limitless possibilities and then mining the road to those very treasures. Some might argue that human beings have no right to inhabit the skies: "If God had meant us to fly he would have given us wings." But a statement such as this is nothing more than a denial of the human spirit.

It is certainly the case, however, that we are not built for flight. Our bodies are fragile, simply not equipped to cope with impact. We can neither glide nor bounce. We are incapable of breathing at high altitudes and die quickly in the freezing temperatures that are to be found just a few thousand feet above the surface of the Earth. And yet we have learned to fly. The pursuit of this goal, arguably the greatest achievement of humankind, has been at the cost of thousands of human lives, many of which were lost during the age of the great airships.

In most respects, the airship represents a tragedy simply waiting to happen. Even given the disturbing number of deaths that occurred during the heyday of the airship, it is surprising that more people did not die. An airship is really just a huge, perilously thin bag of explosively inflammable gas driven along by a sparking engine. It would be hard to imagine a more dangerous combination unless, of course, one tied a small basket to this huge incendiary device and foolishly allowed a few people to climb on board. Nobody set out to endanger lives, it just happened to be the only way that these remarkable craft could be built given the age and the available technology.

Hydrogen, the lightest of all gases, seemed to be the obvious choice as the most efficient means of lifting the airship off the ground. It is plentiful, relatively easy to produce and very cheap. During the early 1920s, for example, a thousand cubic feet of hydrogen could be bought for just a couple of dollars. Unfortunately, for anyone caught anywhere near a hydrogen explosion, the gas combines rather over-enthusiastically with oxygen the moment it is exposed to so much as a spark.

Helium, on the other hand, being an inert gas, simply does not burn. But despite being one of the most abundant elements in the universe, helium is

THE DEVELOPMENT OF OIL WELLS IN TEXAS LED TO THE COMMERCIAL AVAILABILITY OF HELIUM – A FAR SAFER ALTERNATIVE TO THE HYDROGEN COMMONLY USED IN AIRSHIPS.

not easily come by on our planet. It was not isolated until 1895 and became commercially available only as a by-product of the Texas oil fields that were springing up at the start of the twentieth century. A measure of its relative rarity can be gauged from the fact that when hydrogen was being sold for just $2 per thousand cubic feet, an identical volume of helium cost as much as $100 more. Combined with the high cost of helium was the fact that it provides less lift than hydrogen. Considering that many of the airships built in the early 1920s needed to be filled with several million cubic feet of gas in order to get off the ground, only the most safety-conscious airship operator would have chosen to fill his craft with helium.

The age of the airships was also a period during which engineers were making great leaps forward in terms of our understanding of the stresses involved in flight. Aeronautical engineering was in its infancy and so it was inevitable that mistakes were made in pursuit of the all-important perfect weight-to-lift ratio.

Allied to the dangers posed by explosive gases and imperfect engineering were the inevitable games of chance any traveller is forced to play with the weather. A well-engineered helium-filled airship could still find itself being swatted like a fly by the advance of a storm front – one of nature's many random acts of violence.

Just such a case was that of the *Shenandoah*. The *Shenandoah* represented an American dream. It was the first large rigid airship to be built in the country and was designed to be stronger and safer than any airship that had preceded it. To this end, the designers made numerous alterations to the structure of the airship and, for the first time in history, decided that it would be filled with helium. This was done partly to reassure a nervous American public after the outcry that followed the crash of the Italian-built, US Army airship *Roma*, which had burst into flames after colliding with power lines.

THE AMERICAN AIRSHIP *SHENANDOAH* WAS SHATTERED BY A VIOLENT STORM AND BROKE IN TWO BEFORE FALLING TO THE GROUND.

Bearing in mind all of the effort that was put into making the

Shenandoah the safest airship in the skies, it was doubly unfortunate that it should meet its end not in an explosion, but in a freak storm the like of which no airship could be built to withstand.

CROWDS FLOCKED TO SEE THE WRECK OF THE *SHENANDOAH* THE DAY AFTER ITS CRASH.

Very early in the morning of September 3, 1925, the airship was caught, without warning, by an approaching weather front. Even with the engines running at full power, the *Shenandoah* could make no headway in the fearsome wind that was blowing straight at it. Suddenly the *Shenandoah* began to rise uncontrollably, levelling off briefly at around 3,000 feet, a fraction below the limit of its safe travelling height. Just when it looked as though the worst might be over, the airship was caught again, this time in a far more violent current of rising air. Faced with the prospect of the pressurized helium bags bursting in the rapidly thinning air, the captain ordered the crew to open the valves and release some of the precious lifting gas. As the airship began to drop rapidly, the crew released some water ballast to counter the fall, but by now the *Shenandoah* was truly out of control.

The released ballast had halted the drop briefly but once again the airship began to rise, this time nose first. While in this vulnerable position, the *Shenandoah* received a full punch to the body from a sudden massive gust of wind and was smashed to pieces. For a moment a rear section of the ship plunged towards the ground before levelling off and floating slowly down to Earth, much to the relief of the 22 men trapped in that section. For the men trapped in the front part of the airship, however, the ordeal was far from over.

Now free of the weight of the rest of the ship, the nose section began to climb skywards with seven men still on board, holding on for dear life. On and on the climb continued until, at 10,000 feet, it came to an abrupt halt. Showing remarkable skill and courage, the remaining crew began gradually to release helium from the forward section and eventually managed to bring it down from 10,000 feet for a controlled landing. After half an hour of living hell, the ordeal was finally over. America had put its strongest airship up against the weather and been found wanting. Incredibly, 29 of the original 43 crew survived the night to tell the tale. Next time they would not be so lucky.

But America was not alone in its battles with the weather. In 1928, two years after Norwegian explorer Roald Amundsen and the Italian engineer Umberto Nobile had first reached the North Pole by airship,

THE *ITALIA*, DESIGNED BY UMBERTO NOBILE, WAS INTENDED TO HELP SCIENTISTS MAP THE ARCTIC FROM THE AIR.

THE *ITALIA* FLYING OVER SPITSBERGEN BEFORE ITS ILL-FATED VOYAGE ACROSS THE ARCTIC.

Nobile returned with a new airship, the *Italia*, and a handful of scientists. Their plan was to attempt to map the one and a half million square miles of unexplored, icy wilderness that made up the Arctic wasteland and gather as much scientific information as they could during the course of a number of flights across the frozen landscape.

The first flight passed without incident, but disaster struck on the second mission. Without warning, the airship lost buoyancy and crashed down on to the ice below. Nobile and nine of the crew were thrown from the ship by the impact, along with some basic provisions and, as luck would have it, the radio. Of the nine men who tumbled out on to the ice with Nobile, one was killed immediately and another suffered a broken leg. Nobile himself suffered a broken arm and leg in the crash.

Unfortunately, with ten men, the radio and some provisions now scattered over an ice floe, the airship became buoyant again and began to drift skyward with six men still on board. Despite their best efforts, the remaining crew were unable to bring the *Italia* under control and began to drift off into the freezing Arctic fog. Neither the six crew members nor the *Italia* were ever seen again.

Regular distress calls were sent by the surviving radio operator but went unheeded because the *Italia*'s support team had not bothered to

BARNES WALLACE, DESIGNER OF BRITAIN'S *R100* AIRSHIP.

monitor broadcasts. (They had already assumed that everyone must be dead when the ship failed to return to base.) The men remained trapped on the drifting ice floe for nine days until finally, on June 6, 1928, their distress calls were heard. Supplies were dropped to the men by aeroplane on June 20, but the rescue mission continued to be hampered by bad luck and breathtaking incompetence. Remarkably, the last of the stranded men was not removed from the ice floe until July 12.

At the time the Italians were attempting to explore the frozen Arctic, Britain, keen to regain its position as a major player in world affairs, had its own ambitious plans for the airship. With the end of the First World War it soon became evident that Great Britain was no longer the power it had been. Although the empire had remained more or less intact, the mother country was no longer considered to be quite as mighty as it once was. During the 1920s, it became clear that Germany was, like its airships, on the rise once again. The military donkeys who had led so many brave British lions to their deaths in the trenches and on the battlefields of Europe decided that something must be done to stop the rot. Their plan was to use the new technology to restore Britannia to her rightful place as the world's leader. To this end, it was proposed that two airships be built. One of these, the *R100*, was designed by Barnes Wallace. (Wallace was an engineering genius whose career suffered as a result of his uncontrollable ability to make certain sections of the British government feel inadequate in the presence of a man of his intellect.) The other airship was to be the *R101*.

Working on the premise that bigger is better when it comes to making jingoistic gestures, the *R101* was designed to be the largest airship of its day. Unlike the under-funded *R100*, no expense was spared in the production of the *R101*. With little regard for any engineering considerations, a large lounge, dining room and overly luxurious cabins were added to the design along with an asbestos-lined smoking room.

The *R100* takes to the skies on its maiden flight.

A certain amount of stainless steel was also included in the rigid frame of the airship; this is stronger than the Duralumin that was normally used for building the frame but unfortunately much heavier.

As soon as someone bothered to check the mathematics of the airship's design, it became clear that the *R101* would be doing well to get off the ground. A refit was ordered, during which the frame of the airship was split in half and extra sections added in order to provide more lift. Where possible, any unnecessary weight was trimmed to a minimum and the botched airship was soon being prepared for its maiden flight to India – despite having been put through barely any flight testing.

THE *R101* SEEN OVER BEDFORD IN 1929, AFTER LEAVING CARDINGTON ON ONE OF HER TEST FLIGHTS. SHE SUFFERED FROM SEVERAL TECHNICAL PROBLEMS, INCLUDING ENGINE DEFECTS.

On October 4, 1930, a procession of the great and the good braved the driving wind and rain and boarded the *R101* ready to set off for their ultimate destination of Karachi. Flying dangerously low because of its excessive weight, the *R101* headed towards France and into 50-mile-an-hour winds and constant rain.

The airship eventually reached the continent but was by now way behind schedule. Unknown to the passengers on board, there had been an earlier problem with the fabric that had been used to cover the airship. In an attempt to reinforce the fabric, extra patches of material had been glued on using a solution of rubber. This had reacted badly with the doping agent that had already been used to coat the fabric, causing it to become brittle and flaky – two of the worst properties one could wish for in the flexible outer-covering of an airship. Although most of the substandard covering had been replaced, two key sections which had been treated with the rubber solution remained in place. In an act of homicidal stupidity, the decision had been made to leave the sections as they were rather than replace them and risk cancelling the by now heavily publicized flight.

As the low-flying airship cruised through the skies near Beauvais, northern France, the front section of the fabric covering the airship split apart, exposing the hydrogen gas bags underneath. These quickly began to deflate, causing the nose to dive sharply and rendering the airship almost uncontrollable. With no margin for safety, the *R101* ploughed straight into the side of a hill. The smoking room, so long a proud feature of the airship in the minds of its designers, contained the necessary spark that ignited the leaking hydrogen gas. By the time the flames and choking fumes had done their work, 48 passengers and crew lay dead on the French hillside.

The fallout from the accident, if it can be called that, was monumental. The British government ordered that the *R100*, still in its hangar for a refit, be dismantled. They also cancelled plans for two even larger airships, the *R102* and *R103*. As far as Great Britain was concerned, the age of the airship had reached its tragic conclusion.

Unprecedented crowds turned out to pay their respects to the 48 coffins that lay in state at Westminster Abbey in London. (None of the bodies was on show as they had all been burned beyond recognition.)

POLICE AND FIREMEN SEARCH FOR BODIES IN THE WRECKAGE OF THE *R101* ON A HILLSIDE IN NORTHERN FRANCE.

For a moment the Akron reared up at an angle of 45 degrees before the entire stricken airship slapped down into the boiling seas.

Among those in attendance was Hugo Eckener and a small group from the Zeppelin Company, who could find solace in the sure and certain knowledge that their airships were of superior design, and could take comfort from the fact that not a single life had been lost in a German airship disaster.

But the Italians and British were not alone when it came to major airship disasters. Over in the United States, things were hardly any better. On the night of April 3, 1933, the officers and crew of the US craft *Akron*, a navy airship and the largest in the world at the time, set off over the Atlantic Ocean and into the worst storm in living memory. The ship managed to ride out the storm, despite being battered from all sides, until shortly after midnight, when finally the storm proved too much for the *Akron*.

Constant roaring winds caused the rudder control cables to snap. Despite dropping all available ballast, the front of the airship shot upwards. A faulty altimeter had caused the pilot to believe that he was at least 800 feet above the water when, in fact, he was barely clearing the surface of the sea. As the nose of the *Akron* pitched upwards, its tail was dragged through the icy water below. For a moment the *Akron* reared up at an angle of 45 degrees before the entire stricken airship slapped down into the boiling seas.

The Atlantic Ocean during a storm is a magnificent sight, but it is hard to imagine a worse place to be without a life-raft. The crash had happened so quickly that there was no time to launch the life-rafts. Although the alarm had been sounded moments before impact, most of the crew had been asleep. Of the 76 officers and men who had set out on patrol that night only three survived until the morning. For days afterwards, bodies and wreckage drifted ashore all along the New Jersey coastline. This had been the worst disaster in the history of aviation. But history has a habit of repeating itself. Just two years later there occurred an incident which was to bring to an end America's love affair with the airship.

The *Macon* was the undisputed pride of the US Navy. Thanks to a remarkable hook-and-crane system, the *Macon* was able to act as an aircraft carrier, sending out small biplanes to act as scouts. After completing their mission, the aeroplanes would return to the mothership and be plucked out of the sky on the end of a hook which was attached to a crane on board the *Macon*. As soon as the little aeroplanes were captured in this way they would be winched back into the body of the airship and stored away until they were next needed. This was a daring operation which called for great skill on the part of all concerned, but proved its worth during many training missions over the sea.

Although the airship was designed primarily for use over water, it had been used over land – West Texas – back in 1934. The *Macon* had

THE UNITED STATES AIRSHIP *MACON* UNDER CONSTRUCTION IN ITS LOFTY HANGAR.

THE *MACON* FLIES OVER SAN FRANCISCO IN 1933; TWO YEARS LATER IT PLUNGED TO ITS END IN THE OCEAN.

encountered high winds during this training mission and suffered damage to its tail section. It was clear that the airship was always going to be prone to damage in this area until the tail section could be strengthened. Several people argued, however, that the airship was unlikely ever to encounter such severe weather again and so alterations were delayed until such time as it would be convenient. In short, the work was never carried out.

On the evening of February 12, 1935, the *Macon* was returning to base after a training mission when it was hit by a sudden gust of wind just as it was making a turn. It was immediately clear that the tail section had been compromised by the blast. A sudden loss of pressure in the rear section caused the airship to pitch upwards sharply. The captain of the ship immediately ordered that all ballast be dropped, but rather than correcting the fault, this simply caused the airship to rise uncontrollably into the sky.

Before anyone knew what had happened, the stricken airship shot up to 5,000 feet, blowing the automatic valves on all the remaining gas bags. The *Macon* hung motionless at 5,000 feet for over 15 minutes before finally plummeting sickeningly and inevitably back down to sea level. Fortunately, although the temporary halt at 5,000 feet had been a terrifying experience for all on board, it had given everyone time to don life-jackets and make ready the life-rafts. The captain had also had time to call for help over the radio. By the time the ship hit the water, help was already on its way. Of the 83 men who began the descent to the sea from 5,000 feet, 81 survived to be rescued. True disaster had been averted, but the American military would never again make use of the large airship. The one remaining airship it possessed, the *Los Angeles*, was decommissioned, never to fly again. Germany now stood alone as the only country in the world with enough faith in the great airships to continue to build them. And what wonderful airships they were.

Chapter 4

Pride of the Reich

The new airship *LZ129*, as yet unnamed, seen from her tail rudder as she rests in her hangar at the Zeppelin works in March 1936.

Any German airship that was to follow the *Graf Zeppelin* would have to be something special. The *Graf Zeppelin* had set new standards in terms of design, comfort, reliability and sheer magnificence.

It had shattered records and, with its round-the-world flight in 1929, had proved to the world that the skies, for the time being at least, belonged to the great airships. But before any of this, in fact before *Graf Zeppelin* had even lifted off the ground, plans were already being drawn up for an even more ambitious craft.

The *LZ128*, had it been built, would have rivalled Britain's *R101*. At over 750 feet long and packed with more than five million cubic feet of hydrogen gas, it would have resembled an overfed version of the *Graf Zeppelin*. But when the *R101* crashed into a French hillside in 1930 and was destroyed in the resulting hydrogen-fuelled inferno, plans for the *LZ128* were abandoned in favour of a new design, which would be filled with helium, an altogether safer lifting gas.

In order to compensate for the reduction of lift caused by filling the gas cells of the new airship with helium rather than hydrogen, it was decided to construct a craft that was longer and larger than any that had come before it. This airship was to be so large that it would require an enormous new shed to be built at the Zeppelin Company's base at Friedrichshafen, in Germany. But the new airship, which had been given the serial number *LZ129*, was to be born into a troubled world.

Under the terms of the Treaty of Versailles, drawn up at the end of the First World War, Germany was required to make reparations to the Allies. Essentially, these were financial penalties imposed on Germany for the damage done during the war. In 1921, the penalty was fixed at 132 billion gold marks. A number of payments were made towards this final bill, but in 1923 the German government announced that it would be handing over no more money.

In response, the French and the Belgians marched into the Ruhr – Germany's industrial heartland based on the River Rhine – and attempted to take control. With the full backing of the German government, industrialists and workers in the Ruhr brought production to a virtual standstill. To add to the chaos, the German treasury began printing enormous quantities of paper money. By 1924 the German mark was almost worthless. What had previously been a moderate, centre-right-based society was beginning to break down. Democracy was soon under threat as the Communists and the National Socialists (Nazis) began vying with each other for control of the country. In the elections of 1930, the Communists won 77 seats and the Nazis acquired 107. The then chancellor, Heinrich Brüning, was unable to

PAUL VON HINDENBURG, PRESIDENT OF GERMANY FROM 1925 TO 1934.

form a majority government and was able to rule only by an emergency decree issued by the president, Paul von Hindenburg.

Chancellor Brüning resigned in May 1932 and in elections held in July of that year the Nazis gained 230 seats. On January 30, 1933, President Paul von Hindenburg appointed Adolf Hitler – the leader of the Nazi party – as chancellor of Germany. This was done in the naive belief that Hitler could be controlled by his social "superiors" – the traditional German elite made up of conservatives, senior army officers and the president himself. Hindenburg could not have been more wrong. Within two years, Hitler had established a totalitarian state and the old order was all but wiped out.

Initially this appeared to be no bad thing for the Zeppelin Company, which was having difficulties raising enough money to ensure that construction of its new airship, the *LZ129*, would be completed. In 1934, Dr Joseph Goebbels, the minister for propaganda under the new

HITLER INSPECTING TROOPS FOLLOWING HIS APPOINTMENT AS CHANCELLOR.

THE FRAME OF THE *LZ129* UNDER CONSTRUCTION IN THE ZEPPELIN WORKS AT FRIEDRICHSHAFEN.

Nazi regime, was made aware of just how valuable a symbol of German might the new airship would prove. He publicly pledged his support for the project by awarding over two million marks towards its construction costs. Hermann Goering, Goebbels' political rival, had little time for airships, but saw that his role as air minister was being undermined by the propaganda minister's gesture. Having first contributed nine million marks to the project, he then moved to set up a new airship company. In 1935, the German Zeppelin Transport Company (Deutsche Zeppelin-Reederei) was formed with the support of the state-funded airline Lufthansa, which contributed the equivalent of over one million dollars of operating capital to the project. In return, the Zeppelin Company provided the *Graf Zeppelin*. By selling its soul to the Devil, the Zeppelin Company had seen to it that the future of the airship was now assured. But there may have been some who considered the price to be too high.

WORK ON THE GREAT AIRSHIP NEARS COMPLETION.

Hugo Eckener, who had done so much to rescue the Zeppelin Company from economic collapse at the end of the First World War, made no secret of the distaste with which he viewed the Nazis. He despised Hitler and his thuggish cohorts and was prepared to use his position as a very prominent public figure to attempt to raise support for Chancellor Heinrich Brüning, Hitler's predecessor and political opponent. Eckener went so far as to make a national radio broadcast on Brüning's behalf, but even this could not save the ailing chancellor. When Brüning resigned, Eckener was approached and asked to run for the presidency, in the hope that he might be able to impose some limitations on the growing power of the Nazis. Eckener declined to run for the post, choosing instead to concentrate on his airships, but he continued to criticize the Nazis publicly whenever the opportunity arose.

Hermann Goering, the Air Minister, watching the *Hindenburg*; the Nazis regarded the great airship as a symbol of their power.

Ernst Lehmann, Dr Eckener's right-hand man at the Zeppelin Company, appeared to hold no such views on the Nazis, or if he did, then like so many Germans he kept them to himself. By riding the tide of the new movement, Lehmann was able to secure for himself the job of director of the new German Zeppelin Transport Company. Eckener, his former mentor, was given the splendid title of Honorary Chairman of the Board, but in reality the post was little more than that, an empty title that brought with it little or no influence over the running of the company.

A measure of the degree to which the men had, in effect, swapped places can be gauged by two events that occurred after the Nazis effectively took control of the Zeppelin Company. In March 1936, not long after the *LZ129* first emerged from the giant shed where it had been built, the order came down from on high that it was to be given the name of *Hindenburg*. The significance of naming the airship after

THE ZEPPELIN COMPANY'S ADVERTISEMENT FOR ITS NEW AIRSHIP, THE *LZ129*.

the former German president who had been responsible for appointing Hitler as chancellor of Germany in 1933 was lost on no one. Eckener was given no choice in the matter, and may have chosen to go along with his former protégé's decision not to oppose the Nazis on the grounds that, if nothing else, Hindenburg had been something of an old-style German patriot. But what truly caused a rift between the two men was Lehmann's decision to allow the *Hindenburg* to be used, along with the *Graf Zeppelin*, on a propaganda tour just a few days before a referendum on March 29, 1936.

Both men knew that the *Hindenburg* had not been through the comprehensive series of tests that the *Graf Zeppelin* had endured when Eckener had been in charge. The *Hindenburg* was due to make its maiden flight, to Rio de Janeiro in Brazil, the day after the referendum, but Lehmann had taken the decision to cancel the final trials of the new airship in order to accommodate the wishes of Joseph Goebbels and his propaganda tour. Lehmann roused Eckener's fury further by attempting to launch the *Hindenburg* in strong winds. This was a foolish action which resulted in damage to the tail of the airship and an outpouring of abuse by Eckener that was directed as much at Goebbels as it was at Lehmann. (This incident was to take on greater significance when news of Eckener's tirade reached Goebbels.) But, when all was said and done, Hugo Eckener's dream ship had been built. And what a ship it was.

Although the *Hindenburg* was designed to be filled with helium, and had therefore been made much larger than might otherwise have been the case in order to compensate for loss of buoyancy that resulted from using this less efficient lifting gas, the airship's 16 gas cells were eventually filled with hydrogen. When the airship was being designed, the Helium Control Act of 1927 was still in operation. (The American

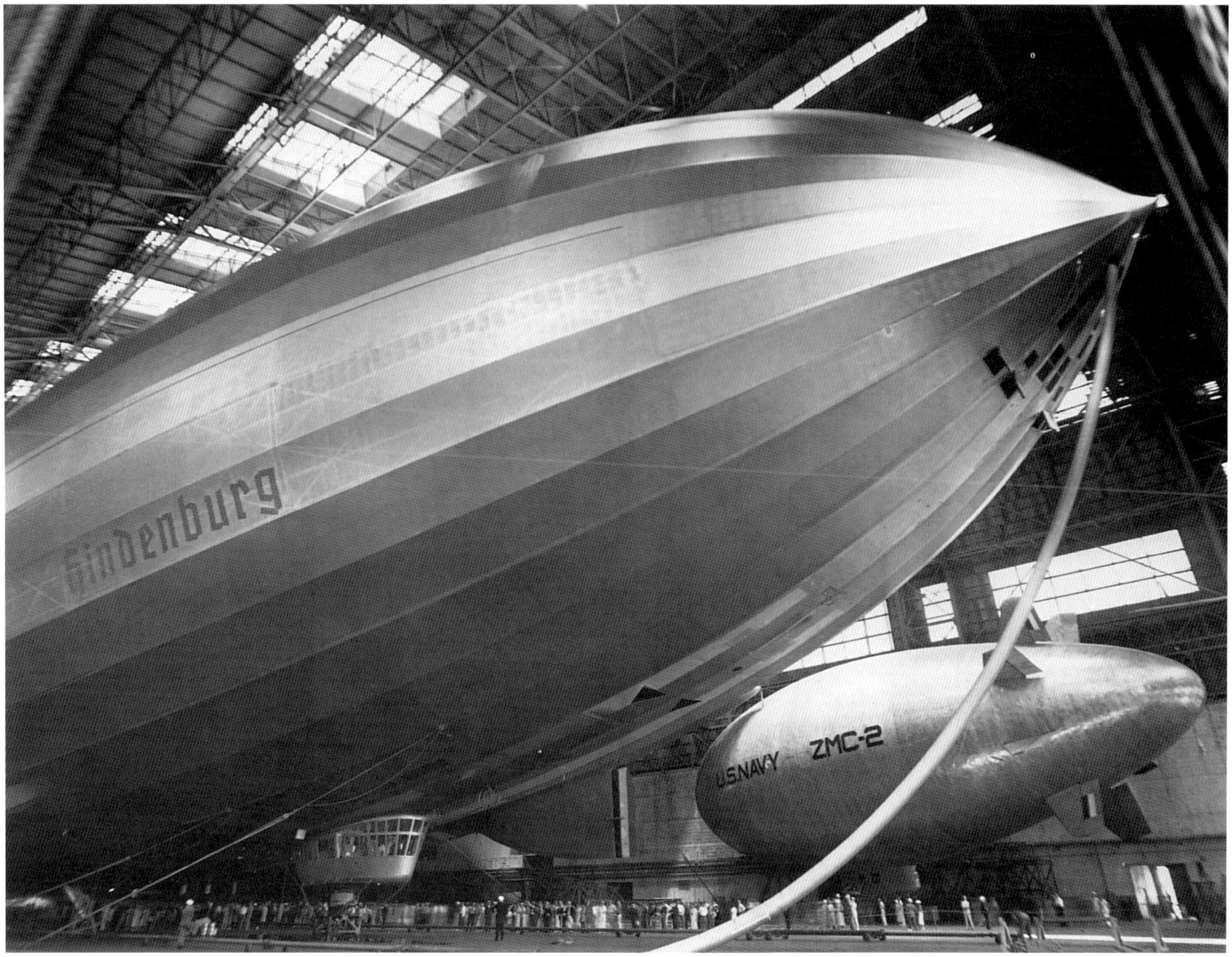

Now named the *Hindenburg*, the airship is seen here in the hangar at Lakehurst.

government had passed the act in order to stop the export of helium for military purposes.) Everyone had assumed that the act would have been repealed by the time the *Hindenburg* took to the air, but this proved not to be the case. Because America held a virtual monopoly on helium, the Germans had no choice but to inflate the *Hindenburg* with hydrogen. At the time, no one could have guessed the tragic consequences of America's refusal to share its helium with the rest of the world. In any case the Zeppelin Company, or the German Zeppelin Transport Company as it had become, appears to have made no effort to persuade the American government to relax its restrictions.

When the *Hindenburg* took to the skies in 1936, it was, and remains, the largest flying machine in history. It is hard to imagine the impact that the *Hindenburg* must have had on all who saw it. At almost 804 feet

A DIAGRAM WAS PUBLISHED TO SHOW THE DETAILS OF THE *HINDENBURG*'S DESIGN, INCLUDING HER LUXURIOUS PASSENGER ACCOMMODATION.

in length, it was just marginally shorter than the *Titanic*. At its widest point, it was a little over 135 feet in diameter. In order to get its enormous but wonderfully elegant bulk off the ground, the *Hindenburg* had a gas capacity of well over seven million cubic feet.

But despite its great size, the airship was, to all intents and purposes, based on traditional designs and constructed using tried-and-tested materials and techniques. (In one of the few breaks with tradition in the design of the *Hindenburg*, the bags which held the hydrogen gas were lined not with the intestinal outer membranes of cattle – known as "goldbeater's skin" – but with a gelatine solution that had first been used by American airship builders.) The girders that were used to give rigidity to the airship were made of Duralumin, a lightweight alloy commonly found in airships at the time. The main frame of the craft was supported using a braced wire technique, which had also been adopted by aeroplane designers for use on the biplane. This technique exploits the inherent strength of triangular-based structures and effectively divides the main frame into a series of tightly wired triangles. Each of these individual triangular structures, which are really little more than wire and air, contributes its own natural rigidity to produce an overall structure that is both incredibly strong and yet remarkably lightweight.

An airship of this size was going to need some very powerful engines if it was to be at all manoeuvrable once in the air, and to this end the *Hindenburg* was fitted with four brand-new diesel engines supplied by the Daimler-Benz Company. These engines, along with the airship's rudders and elevators (which determine the direction and attitude, or angle, of flight) were operated from the control car, which was slung under the nose of the *Hindenburg*, in between the officers' quarters and the radio room. (The *Hindenburg* also had an auxiliary control room in its tail section.)

Fuel, oil and water tanks were located along the length of the airship and the crew's quarters, which were built into the body of the *Hindenburg*, could be found at the front of the craft (behind the passengers' quarters) and also near the tail section. There were also freight rooms, a mail room and, of course, a radio room near the front of the *Hindenburg*.

THE CONTROL CAR LOCATED UNDER THE NOSE OF THE GREAT AIRSHIP.

ALL THE AMENITIES OF HOTEL LIFE IN CLOUDLAND: THE PICTURES SHOW PASSENGERS ENJOYING FACILITIES ON THE *HINDENBURG*, INCLUDING THE VIEW FROM AN OBSERVATION WINDOW (LEFT).

The crew's mess, kitchen, officers' mess, smoking room, toilets and bar were all situated on B deck, just behind the control car but within the body of the craft. There was also a shower fitted on this deck, which was a first for any airship.

Although the *Hindenburg* was truly magnificent to behold, it did not appear, from the outside at least, to be in any sense out of the ordinary (apart, of course, from its size). But anyone who climbed on board the ship as a passenger soon realized that they had entered a whole new world.

Despite being the largest aircraft ever built, the *Hindenburg* catered for only 50 passengers. But those 50 passengers were treated to a degree of luxury and comfort that has yet to be surpassed by any commercial aircraft built since Hugo Eckener's dream ship dominated the skies. The passengers' quarters were located on A deck, just behind the control car and directly above B deck. These quarters housed 25 double bedrooms, each with bunk beds of the type found on sleeper trains. Although these rooms were undeniably almost spartan in comparison to the rest of the facilities on board, the *Hindenburg*'s designers had reasoned, quite correctly, that the passengers would spend most of their time in the public areas of A deck. Because of this, no expense was spared in making that deck as luxurious as possible.

The public areas on A deck were situated on either side of the bedrooms. Both sides of A deck had promenades, which were fitted with picture windows slanted at an angle of 45 degrees in order to give maximum visibility of the sights below. On the port, or left-hand, side of the airship, just behind one of the promenades, could be found the dining room. The meals on offer were the match of anything found in

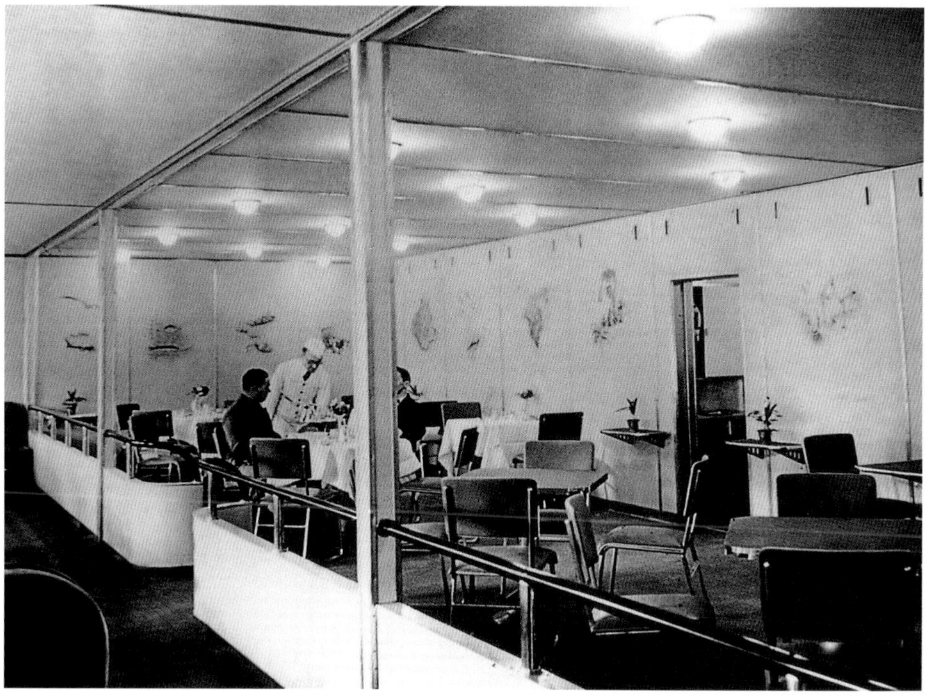

Part of the lounge where passengers could relax on the voyage.

the better restaurants of Europe and passengers could also select from a fairly extensive list of German and French wines.

The lounge was to be found on the starboard, or right-hand, side of the airship. This room was truly remarkable, and was even fitted with a baby grand piano. The piano had been especially constructed using aluminium and weighed less than 400 pounds, making it quite a feat of engineering in itself. Next to the lounge was the reading room, which also doubled as a writing room, with a selection of *Hindenburg* stationery available to passengers, as well as a post box (which was emptied twice daily). A small bar could be found down on B deck, next to the smoking room. The smoking room could be entered only by passing through an airlock, which had been fitted to reduce the risk of any stray hydrogen being ignited by an unwary smoker.

In short, the *Hindenburg* was truly the most remarkable airship ever built. It represented the culmination of one man's dreams, but when it set off on its first Atlantic crossing in 1936, very few of the passengers could have imagined what that dream had cost Dr Hugo Eckener, nor the further indignities he would face as a result of his unholy pact with the Nazis.

One of the promenades, showing the slanting observation windows.

Chapter 5

The Voyages of the Great Ship

Hugo Eckener, who rescued the Zeppelin Company from collapse after the First World War.

On March 31, 1936, the *Hindenburg*, pride of the Third Reich and the largest airship ever to grace the skies, set off for Rio de Janeiro.

The calm and stately manner of the *Hindenburg*'s departure was, however, in stark contrast to the raging storm that was bubbling just under the surface on board the airship. Two days earlier the *Hindenburg* had been on a propaganda tour with her sister ship, the *Graf Zeppelin*, sent out to urge patriotic Germans to vote "yes" in a referendum on the reoccupation of the Rhinelands. Now this symbol of a reborn, confident Germany was to make its first transatlantic crossing.

Had Hugo Eckener had his way, the *Hindenburg* would have been taken out for more tests and trials before such a feat was attempted. Before the Nazis took over his beloved company, Eckener would never have allowed one of his airships to be sent on such a trip without extensive testing. The propaganda tour, undertaken on the ordered

NAZI BROWNSHIRTS, SEEN HERE IN BERLIN IN 1936.

request of Joseph Goebbels and carried out with the collusion of Ernst Lehmann, had torn a hole in the test schedule of the *Hindenburg*. This meant that the newly designed Daimler-Benz engines had not been through the full power trials that Eckener would have insisted on. But now Eckener was in no position to insist on anything.

For the first time since anyone could remember, Hugo Eckener was not in command. The man who had brought the Zeppelin Company back from the brink of collapse was now little more than a figurehead – a familiar, reassuring presence who in reality was just another passenger. And this passenger had a few complaints to make.

Hugo Eckener was the product of a Germany that had died with the man after whom his magnificent airship had been named. The old Germany had been ruled by a Prussian elite who were confident to the point of arrogance and as aristocratic as any noble English dynasty. These men had assumed power not by scheming and murdering, but by virtue of their belief in their God-given right to rule. This regime was by no means perfect, but the principle of honour, above all other things, ran through every layer of the society they created.

What Eckener found so repugnant about the Nazis was not the fact that they were drawn from an underclass who were unfit to rule –

although he almost certainly believed this to be the case – but the fact that they were men without honour. He had seen their leader, a jumped-up corporal from the lower orders, lie and cheat his way into power. He could not have escaped the thuggish presence of Hitler's brownshirts in every city in Germany. These were not the kind of men whom Eckener had been raised to respect. These were little more than gangsters, criminals who demanded respect by means of knives and clubs and guns. And at a time when the Nazi regime was becoming more murderous by the day, Eckener made himself a marked man by refusing to keep his opinions to himself.

He made no secret of his views on Goebbels' decision to order the *Hindenburg* out on the propaganda tour, views which soon became known to Hitler's propaganda minister. Eckener had been especially scathing in his comments about the referendum itself, comments which had once again found their way back to Goebbels. Before the *Hindenburg* had even reached its destination, Eckener was going to find himself in deep trouble with the new regime.

JOSEPH GOEBBELS, HITLER'S PROPAGANDA MINISTER, PHOTOGRAPHED IN HIS STUDY IN 1934.

Despite the lack of proper trials, the journey to Rio de Janeiro passed without any serious problems for the *Hindenburg*. Eckener's journey proved, however, to be far from trouble-free. Halfway across the Atlantic he was approached by a reporter from the London *Times*, who requested his views on the banishment order that Joseph Goebbels had placed upon him. Eckener was unable to answer the reporter as this was the first that he had heard of such a thing.

By the time the *Hindenburg* arrived in Brazil, the situation was becoming clearer. In his absence, Goebbels had, in effect, declared that Eckener was an enemy of the people. The declaration of banishment that followed made it clear that neither Eckener's name nor photograph was to appear in any newspaper in Germany. At a banquet held by the German Club of Rio, Eckener felt compelled to make a speech in which he confirmed that he had been declared a "non-person". He went on to add that he did not in any way share the beliefs of the Nazis and regretted the "tensions and divisions" that they had caused.

That Eckener felt compelled to make this speech is understandable. It was bad enough that these frightful individuals had hijacked his beloved Germany. For them to go further and declare that he was no longer welcome in the country in which he had been born was more than the patrician Eckener could tolerate. But to make such a speech, in the certain knowledge that it would reach the ears of his tormentors, was indeed a brave act. Few people in Eckener's position in Germany could have had any doubts about the murderous intentions of the Nazis. They had proved over and over again that they were prepared to resort to any means in pursuit of their aims. Eckener must have known that one old man, no matter how venerable, was going to prove no obstacle to the Nazis, and yet despite his precarious position, his sense of honour drove him to speak out. As the *Hindenburg* lifted off for the return journey to Germany, Eckener must have wondered what would be waiting for him on his return, although there must have been times on the journey home when he wondered if anyone was going to make it back.

During the outbound journey to Brazil, one of the four diesel engines had broken down. Although temporary repairs had been made, it was not possible to run the engine at full speed. On the return journey, another engine broke down, this time while the *Hindenburg* was off the coast of Africa. While that engine was still in pieces, yet another engine failed. The *Hindenburg* was now in a perilous position. With only one fully

ERNST LEHMANN, ECKENER'S DEPUTY, BECAME DIRECTOR OF THE GERMAN ZEPPELIN TRANSPORT COMPANY WHEN THE STATE SET UP ITS NEW AIRSHIP COMPANY.

THOUGH GOERING WAS UNIMPRESSED BY AIRSHIPS, HE RECOGNIZED THE PROPAGANDA VALUE OF THE MIGHTY *GRAF ZEPPELIN* AND ITS SISTER THE *HINDENBURG*.

functional engine and another that was operating below par, the airship would run into serious difficulties if it met a storm. To the right lay the Sahara Desert, to the left there was nothing but water, and there were no guarantees that the remaining good engine would last another hour.

The three malfunctioning engines had all failed because of a basic design fault. This problem would have been easy to sort out had it occurred during the test flights that Eckener had demanded, but out by the Sahara Desert, Daimler-Benz engineers were hard to find. Ernst Lehmann, who was at the helm, had little choice but to order that the *Hindenburg* climb up in search of the counter-trade winds and attempt to coast home on these. Luck was with Lehmann that day and he found the winds at a little over three-and-a-half thousand feet. The *Hindenburg* made its way slowly back to Germany, proving that Eckener's concerns about his dream airship were well-founded.

The failure of the *Hindenburg*'s engines was, however, to be a cloud with a silver lining for Hugo Eckener. Dr Eckener had returned to Germany to face an uncertain future. The Nazis were by now assassinating opponents and even a man of Eckener's standing could not be certain that the next bullet would not be for him. In a move that would allow all sides to save face, Eckener was approached by the air minister, Hermann Goering, who suggested that he write to Joseph Goebbels and explain that he had objected to the propaganda flight on purely technical grounds. The fact that all of Eckener's fears concerning the airworthiness of the *Hindenburg* had proved well-founded on the journey home would lend credibility to his statement. Eckener, who was initially reluctant to co-operate, eventually agreed to write the letter. Goebbels received Eckener's written explanation with all the grace that is the hallmark of the truly small-minded, but for now Dr Eckener was saved. When, on

May 6, 1936, the *Hindenburg* took to the air at the start of its first trip to the United States, it did so with extensively overhauled engines, which were never again to prove troublesome.

The May 6 flight was of special significance in the history of the airship for here, at last, began the first-ever scheduled air service between Europe and North America. This had long been a dream of the great airship builders and it must have been an especially proud moment for the old guard at the former Zeppelin Company. The airfield

THE *HINDENBURG* ARRIVING AT LAKEHURST IN MAY 1936 ON ONE OF ITS EARLY TRIPS TO THE USA.

Hindenburg

at Friedrichshafen, the traditional home of the zeppelin, was packed with reporters, photographers, guests, ground crew and sightseers. Even the local brass band got in on the act, providing a range of triumphant tunes to stir even the sternest Teutonic heart.

As the sun began to set over the historic airfield, the airship floated slowly skyward and its four big engines kicked into life to propel the *Hindenburg* towards the New World. The flight of the *Hindenburg* was so smooth that several passengers were initially not at all convinced that the great airship had even left the ground. The names of each of the passengers appeared on a list, along with the names of the *Hindenburg*'s officers, which had been distributed just after take-off. The passengers were made up largely from that group which would today be called jet-setters, being a mixture of European aristocracy and media figures, along with a few relatively lowly publishing types and journalists. At the head of the list of passengers and officers was the name of Dr Hugo Eckener, but there was no indication of his position or role on board the airship. Immediately below his name was that of Captain Ernst Lehmann, along with the rather grand title of "Commander". Clearly, Eckener was no longer in favour, but if he minded this fact he certainly didn't show it. For now he seemed content to wander about his wonderful ship and graciously accept the praise heaped upon him for his efforts by his fellow passengers, most of whom had been more than happy to part with the $400 fare for the one-way trip to Lakehurst, New Jersey.

The journey across the Atlantic passed without incident and, a little over 60 hours after setting off from Germany, the great airship landed at 6.10 a.m. at Lakehurst. This brought to an end the first of ten trips to the United States that the *Hindenburg* would make before the winter of 1936 set in and prevented transatlantic airship travel until the following spring. In its first year of flight, the *Hindenburg* carried a total of 1,600 passengers across the Atlantic, travelling a total of 200,000 miles. Aside from a couple of minor brushes with danger, during which the airship almost dipped into the sea and narrowly missed a collision while flying across Newfoundland through freezing fog, all of these flights proved to be remarkably free of incident. As if to underline the point about the safety of travel in his airship, Hugo Eckener set up what became known as the Millionaires' Flight.

Just before returning to Germany for the last time before winter, Eckener invited 72 of America's (and therefore the world's) richest and

As the sun began to set over the historic airfield, the airship floated slowly skyward and its four big engines kicked into life to propel the Hindenburg towards the New World.

Left: People gather at Frankfurt to watch the airship lift off for its journey across the Atlantic.

A coloured drawing by Hans Liska shows the *Hindenburg* sailing over the Reich arena at the opening of the 1936 Olympic Games.

most powerful figures on board for a lavish luncheon during which they were taken for a 10-hour flight. The publicity that the flight generated – sending several billion dollars' worth of multi-millionaires up in an airship is a guaranteed way of grabbing column inches – helped to fix in many people's minds the idea that scheduled transatlantic, and even internal, flights by airship were going to be a very real proposition. The fact that the *Hindenburg* came close to covering its operating costs during this first season made the idea all the more viable. Few who had flown on board the *Hindenburg* could have been in any doubt that the future of passenger aviation lay with the great airships and many eagerly awaited the start of the new season to see just what wonders the former Zeppelin Company had in store for them. But back in Germany, Hitler, Goering and Goebbels had been putting the *Hindenburg* to other uses.

Anyone who sells his soul to the Devil knows that one day the Devil will return to claim what is his. Having provided the funding for so much of the cost of constructing the *Hindenburg*, the Nazis were keen to see a return on their investment. Having first emblazoned their corrupted swastika on the side of the great airship, much to Eckener's disgust, and sent the craft on a propaganda tour with the *Graf Zeppelin*, the Nazis went on to commandeer the *Hindenburg* for the opening of the Berlin Olympics in August 1936.

The Olympic Games were to have been a showcase for the new Germany, and Hitler fully expected that the natural superiority of his Aryan athletes would prove to the world the validity of Nazi ideology. By now the world was becoming increasingly aware of what Nazi ideology really meant for those who were judged, in the opinion of Hitler and his fellow Nazis, to be inferior. Many nations boycotted the event, but in the end it was on the track and field that the Nazis' ideals were dealt their most convincing blow. Jesse Owens, a black American athlete, proved to the world in the best possible way that Hitler's views were unsound: by beating Hitler's Aryan athletes to four gold medals.

Jesse Owens won four gold medals at the 1936 Olympics, thereby discrediting Nazi assertions of Aryan superiority.

Not to be outdone, Hitler then ordered that the *Hindenburg* be sent

on a flypast during the notorious Nuremberg rally in September. As Hitler urged the massed ranks of the militarized party faithful forward on their march to a new German empire, Eckener's dream machine cruised above them, an unmistakable symbol of the might of the new order.

No one who was present that day was in any doubt as to the message that Hitler was sending out to the world: here is our technology, the fruit of superior Aryan minds; as we dominate the skies, so too shall we dominate the world.

THE *HINDENBURG* WAS CONSCRIPTED INTO SERVICE AT THE INFAMOUS NUREMBERG RALLY IN SEPTEMBER 1936.

Chapter 6

The Dream Ends

HUGO ECKENER MEETING PRESIDENT ROOSEVELT IN WASHINGTON, MAY 1936.

No one at the landing site could have guessed that May 6, 1937, was to be a day unlike any other – a day whose memory would remain with them for the rest of their lives.

A year earlier the magnificent *Hindenburg* airship, the pride of the Third Reich, had arrived in America to an ecstatic welcome. So great had been the press interest at the time that the telephone company had been unable to provide enough lines for all the journalists who were covering the event at Lakehurst, New Jersey. Now, apart from a handful of low-grade stringers – local men who had been employed by the big papers to cover what had become a routine story – there was just one newsreel crew and a couple of photographers.

THE *HINDENBURG*'S CLOSE ASSOCIATION WITH THE NAZI PARTY MADE HER A POTENTIAL TARGET FOR THOSE SEEKING REVENGE AGAINST ITS TYRANNICAL RULE.

The *Hindenburg* had begun the new season with the addition of ten extra passenger cabins. The continuing American embargo on the export of helium had meant that the airship had still to be filled with hydrogen. However, the extra lift that the hydrogen provided had been exploited to allow the fitting of the extra cabins. But that wasn't all that was new to the *Hindenburg* in the 1937 season.

THE *HINDENBURG* SAILS LOW OVER THE SKYSCRAPERS OF NEW YORK ON 6TH MAY 1937 ONLY HOURS BEFORE HER FATEFUL LANDING IN NEW JERSEY.

Eckener's airship now had a new captain at the helm, one Max Pruss. The new captain's first job was to take the airship on a round trip to Rio de Janeiro at the start of the season. The trip passed without incident and soon everyone was getting ready for the resumption of the transatlantic passenger service to North America, which was due to start from Germany on May 3, 1937.

A couple of old faces at Deutsche Zeppelin-Reederei, Ernst Lehmann and Hugo Eckener, now also found that their roles were changing. Lehmann had given up command of the *Hindenburg* to concentrate on his real job, that of directing passenger operations and

taking care of the training of airship crews. Poor Hugo Eckener, the man whose dreams were rapidly being turned into nightmares by the Nazis, found himself increasingly sidelined. Having done more than anyone alive to popularize the whole notion of a regular transatlantic airship service, he now found that his own services were no longer required. Although he was later to find himself being sent to America on a mission by the Nazis – who wanted to get their hands on a large supply of helium in preparation for the coming Second World War – Eckener was very much yesterday's man. Despite remaining hugely popular with airship enthusiasts the world over, his clash with the Nazis ultimately dealt a fatal blow to his airship career and he was destined to see out his days in a machine factory.

To anyone boarding the *Hindenburg* at Frankfurt for the crossing to North America on May 3, things seemed pretty much as normal. Aside from the extra passenger cabins, the airship remained largely unchanged. There were far fewer passengers than one might normally expect for such a trip – only 36 rather than the 72 that the *Hindenburg* was now equipped to carry – but the growing political instability in Germany, triggered by persistent rumours about a possible occupation of Austria by Hitler's forces, had kept many people at home.

There was also the additional problem of bomb threats. By virtue of Hitler's shameless use of the *Hindenburg* as a tool of propaganda, the airship was now firmly viewed as a symbol of the growing tyranny of the Nazi party. Many people had suffered under the Nazi regime, although many more were to suffer unimaginable horrors before the Nazis were through, and the *Hindenburg* was an obvious target for anyone seeking revenge. Security was tight but discreet at the new airship hangar at Frankfurt but the possibility that someone might wish to destroy the *Hindenburg* was playing on the minds of all the crew as they made ready to lift off on May 3.

The Atlantic crossing passed without incident. Indeed, there were some on board the *Hindenburg* who were almost disappointed at the lack of excitement during the crossing. The weather had been miserable for the entire trip, but it was not of the storms and gales variety that might interest the more adventurous traveller. This was grey weather: dull, cloudy and damp. There was also a persistent headwind which had been slowing down the progress of the mighty airship for some time and, as the *Hindenburg* approached the coast of North America, the wind was beginning to pick up.

The *Hindenburg* had been due to arrive at the landing site in Lakehurst, New Jersey, at 6 a.m., but the chances of arriving at that time were looking increasingly unlikely. Reluctantly, Captain Pruss radioed ahead to warn Lakehurst that, at current speeds, he might be up to 12 hours late.

Despite the increasingly bad weather, no one on board the *Hindenburg* seemed especially worried. This may have been due to the overwhelming presence of so many experienced airship men among the crew. In addition to Captain Pruss, there was also Ernst Lehmann, who was flying as an observer on Pruss' first North American trip as captain, as well as three other airship captains: Captain Anton Wittemann, who was also acting as an observer; Captain Albert Sammt, in the role of first officer; and Captain Heinrich Bauer, who was in the role of second officer.

Any annoyance felt by the passengers at the late arrival of the airship soon dissipated after they caught their first glimpse of an American city, Boston, through a break in the cloud cover. By now the ground crew at Lakehurst were beginning to get ready for the arrival of the *Hindenburg*. The latest revision of the arrival time suggested that the airship would be docking at around 4 p.m.

The weather cleared up a little as the day went on and as 3 p.m. approached, Captain Pruss decided to take the *Hindenburg* in for a low pass over the city of New York. While the passengers enjoyed unique and unparalleled views of the city's famous sights, Captain Pruss informed the base commander at Lakehurst, Charles Rosendahl, that he wished to make up for time lost to the headwind by returning to Frankfurt at the earliest opportunity – ideally before midnight. Rosendahl understood the captain's desire to make up for lost time, but knew that the storm front that was approaching Lakehurst would almost certainly make a return trip impossible before the following day.

At last, after what seemed like an eternity of waiting, the magnificent airship finally floated into view over Lakehurst just before 4 p.m. A cheer went up from the small crowd, people who were eagerly awaiting the arrival of loved ones or simply waiting to board the *Hindenburg* for the return trip. (Many of those who were intending to join the *Hindenburg* at Lakehurst were on their way to Europe for the coronation of King George VI, who was replacing the recently abdicated Edward.)

Just at that moment a thunderstorm opened up, splitting the heavens with a roar that seemed to shake the very ground beneath the waiting crowd's feet. People ran for cover from the pouring rain and

MANY OF THOSE WAITING TO JOIN THE *HINDENBURG*'S RETURN TRIP FROM LAKEHURST WERE ON THEIR WAY TO SEE THE CORONATION OF GEORGE VI IN LONDON.

Finally the motors of the great ship fell silent and for a moment everyone on the ground held their breath ...

watched from their shelters as the airship moved away from the mooring mast and back out over open country. With winds now gusting at over 25 knots, Captain Max Pruss decided to postpone any attempt at landing. He knew that the storm would make his ship difficult to handle and did not dare to risk a landing under such atrocious conditions.

Captain Pruss knew only too well the dangers of such an exercise. He had learned to respect and, to some extent, fear the ship that was now under his control. Since the turn of the century, the French, the British and the Americans had all suffered tremendous losses as a result of airship disasters. The French had lost one in a mid-air explosion, the British had seen two of theirs crash and burn, and the Americans had lost all three of their giant airships in horrifying crashes. Germany was alone in being the only country in the world to have a perfect safety record for giant airships. Not a single one of their graceful, stately craft had been lost in an accident. Even during wartime, and despite the continued use of the highly explosive hydrogen gas, not a single life had been lost in a German airship accident. Captain Pruss was determined that he was not going to be the first German officer to lose an airship.

Deciding to wait out the storm until the weather cleared up, Captain Pruss headed off on a south-easterly course before turning around and finally making his way back to Lakehurst in the hope that it would be safe for him to land. At just before 6.15 p.m. he received word from Commander Rosendahl that conditions were suitable for landing. By now the *Hindenburg* was some 14 miles to the south of Lakehurst, and Commander Rosendahl seemed keener than ever to get the airship down on the ground and the passengers off the craft while the opportunity presented itself.

By the time the *Hindenburg* arrived back at Lakehurst, the people on the ground were becoming very impatient. Most had been waiting since before the scheduled 6 a.m. arrival time and it was now past 7 o'clock in the evening. The appalling weather had done nothing to improve anyone's mood, which by now was becoming positively hostile. The ground crew, keen to stay away from the complaints of the waiting passengers, set about the task of preparing for the landing with the enthusiasm of men who knew that they could finally go home once the ship had been safely moored.

Everyone on board the *Hindenburg* seemed to be caught up in the excitement of the landing, not least the crew. For some reason, on-board instruments seemed to indicate that the airship was tail heavy by about a tonne. Despite dropping over a tonne of water ballast – some of it on the already soaked spectators below – the tail section remained too heavy and the imbalance had to be corrected by sending crewmen to the nose of the ship, where they could counterbalance the excess weight.

For the next 20 minutes or so, the *Hindenburg* was manoeuvred gracefully into position by the mast, her heavy mooring lines by now trailing along the ground. The weather had improved enormously and the earlier storm was by now reduced to a faint drizzle and a gentle but steady breeze.

Finally the motors of the great ship fell silent and for a moment everyone on the ground held their breath, content to gaze in awe at the wonder that seemed to hang in the air above them as if in defiance of the laws of nature. But nature was not to be overcome.

THE CROWDS CROUCH IN DISBELIEF AS THE AIRSHIP EXPLODES ABOVE THEIR HEADS.

A SEQUENCE OF PHOTOGRAPHS TAKEN AS THE *HINDENBURG* MET HER CATASTROPIC END, ALL OF WHICH HAPPENED IN JUST 34 SECONDS.

It began as a spark, a simple jolt of static electricity that found a companion in the small gas leak that had opened at the top of the hydrogen-filled ship. They danced on the surface of the ship for a second before the union of spark and gas exploded into life, transforming in an instant to become the very deadliest of fires. The ship, which had been floating so serenely, gave the slightest of judders. The crew exchanged fearful glances, they all knew what it was but none was willing to speak the word.

In seconds the fire spread halfway through the ship. So bright was the inferno that the ground below appeared as brightly lit as if it were a summer's afternoon. Confused passengers, who were still waving to their loved ones on the ground below, almost had time to wonder why the people on the ground were now running for cover before the *Hindenburg* pitched violently upwards.

The panic on board was total and absolute. All were faced with the unimaginably terrible choice of burning with the ship or facing certain death by jumping to the ground beneath them. Suddenly fate and nature made the decision for them.

The ship hit the ground tail first with a sickening crunch. Molten aluminium splashed outward from the point of impact as the rest of the ship hit the ground. Incredibly, people began to emerge from the inferno. Some managed just a few steps before dying, their bodies burned beyond recognition. Others had survived the jump, only to be consumed in the flames of the burning wreckage, unable to escape on broken legs and backs. From spark to crash, the whole incident had lasted just 34 seconds.

In that half minute or so, the fate of the airship industry as a whole

The panic on board was total and absolute.

As the airship hit the ground in a mass of flames and smoke, survivors tried to flee from the wreckage.

was sealed. As word of the disaster spread, accompanied by the horrifying newsreel footage, interest in airship flight dried up. Within two years, most of the routes that had been covered by the mighty airships were in the hands of the new aeroplane companies, a situation that was never to be reversed. What had started as the most wonderful of dreams had ended in a nightmare of torn and twisted metal and charred and wasted lives. Succeeding generations were to be denied the awe-inspiring, breath-snatching opportunity of watching a magnificent airship floating steadily, sedately and elegantly into view. These giant cruise liners of the skies were to be no more, consigned like so many dinosaurs to the hallowed halls of museums and the pages of books, to be gazed at in wonder by those who can only imagine, as they would the flight of a pterodactyl, a time when these magnificent birds once dominated the skies.

LEFT: THE FOLLOWING DAY ALL THAT WAS LEFT WAS THE BURNED-OUT WRECKAGE.

The morning of May 7, 1937, was truly miserable. As dawn broke over Lakehurst Field the full extent of the horror that had occurred the previous evening could be seen at last. Just 700 feet from the mooring mast lay the burned-out wreckage of the *Hindenburg*, by now reduced to a bare skeleton of blackened metal. Heinrich Bauer, second officer on the *Hindenburg*, was at the crash site at first light. Being the most senior officer who was still on his feet, he felt compelled to go over the wreckage in search of clues as to what had caused the crash. Even though the airship had exploded less than 12 hours earlier, there were already dark rumours beginning to spread about the possibility that the *Hindenburg* had been destroyed by sabotage.

Back in Europe, Hugo Eckener was already voicing his concerns that the airship had been brought down by opponents of the Nazis, but he was quickly silenced in no uncertain terms by Hermann Goering, Hitler's air minister. Goering and Hitler were determined to stamp out even a hint of a suggestion that there might be factions within Germany who opposed the Nazis. As long as the rumours of terrorist action persisted, it was going to be difficult to present the image of a united Germany to an increasingly sceptical world. Rather than be faced with the indisputable evidence of sabotage that the destruction of the *Hindenburg*'s sister ship would have brought, they ordered the *Graf Zeppelin* to be grounded on its return from Rio de Janeiro.

Just 700 feet from the mooring mast lay the burned-out wreckage of the Hindenburg, by now reduced to a bare skeleton of blackened metal.

Back in New Jersey, Ernst Lehmann was visited on his deathbed by Commander Rosendahl, a trusted old friend, fellow officer and great

THE WRECKED FRAME OF THE *HINDENBURG* LIES UNDER GUARD WHILE INVESTIGATIONS TAKE PLACE.

advocate of the airship. With his dying words, Lehmann expressed his belief that the *Hindenburg* could have been brought down only by sabotage, in this case a bomb. The bomb theory was lent a certain amount of credence by the fact that the airship's arrival had been delayed for over 13 hours. Anyone who had set a bomb on the airship would have assumed that they would be on the ground and well away from the *Hindenburg* before it went off. With this in mind, the FBI set about investigating any passenger or crew member who might have had a motive for carrying out the bombing. The fact that its investigation seems to have been carried out in a half-hearted manner has led some theorists to suggest that the Bureau had no interest in uncovering the truth, but this is probably a little unfair. Considering that the bomber, if one existed, must have been on board the airship when it exploded, there was a very real possibility that he or she was dead. The FBI did investigate one particular passenger, however, a comic acrobat called Joseph Späh.

Späh, a German national, had been bringing a dog back from Germany for his children, who lived with their mother on Long Island. Even though passengers were not usually allowed into the body of the airship without an escort from the crew, Späh had made so many trips to see the dog, which was being stored in the hold, that the crew had given up trying to monitor his visits and left him to make his own way to and from the dog. Despite its efforts, however, the FBI was unable to find any convincing evidence to prove the theory that the *Hindenburg* had exploded as a result of a bomb going off just before landing.

With a lack of any obvious evidence for the cause of the explosion, a team of investigators made up of German and American experts and chaired by Colonel South Trimble Jr was set up to look into the matter. The Germans sent over Hugo Eckener, Ludwig Dürr (who had been the chief designer of the *Hindenburg*), and an expert in the field of electrostatics, Professor Max Dieckmann.

Dieckmann's presence was a strong indicator of the direction the investigation was taking. With the Germans unwilling even to

The rear part of the airship exploded into flames ... then, within seconds, the fire spread along its length.

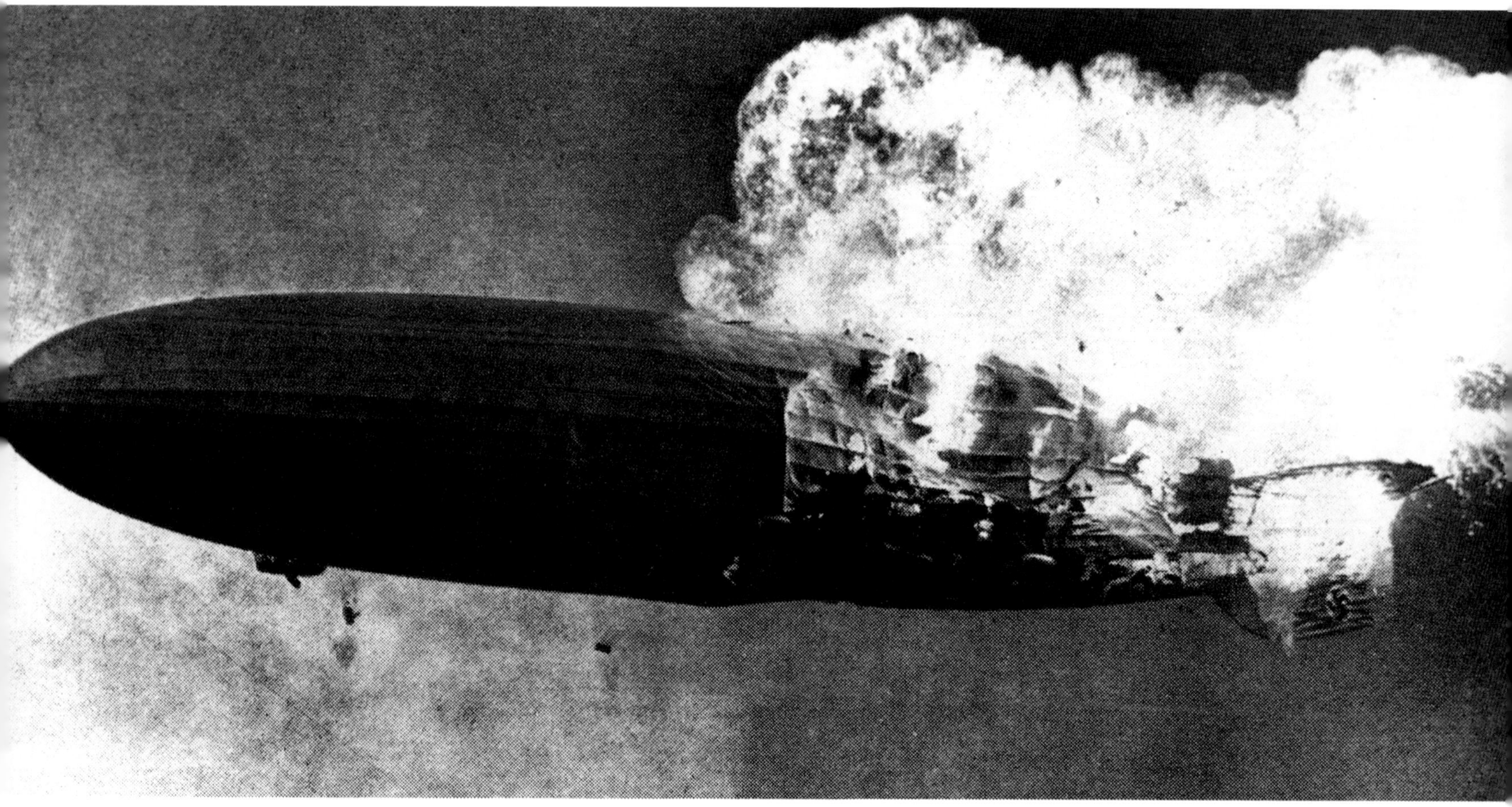

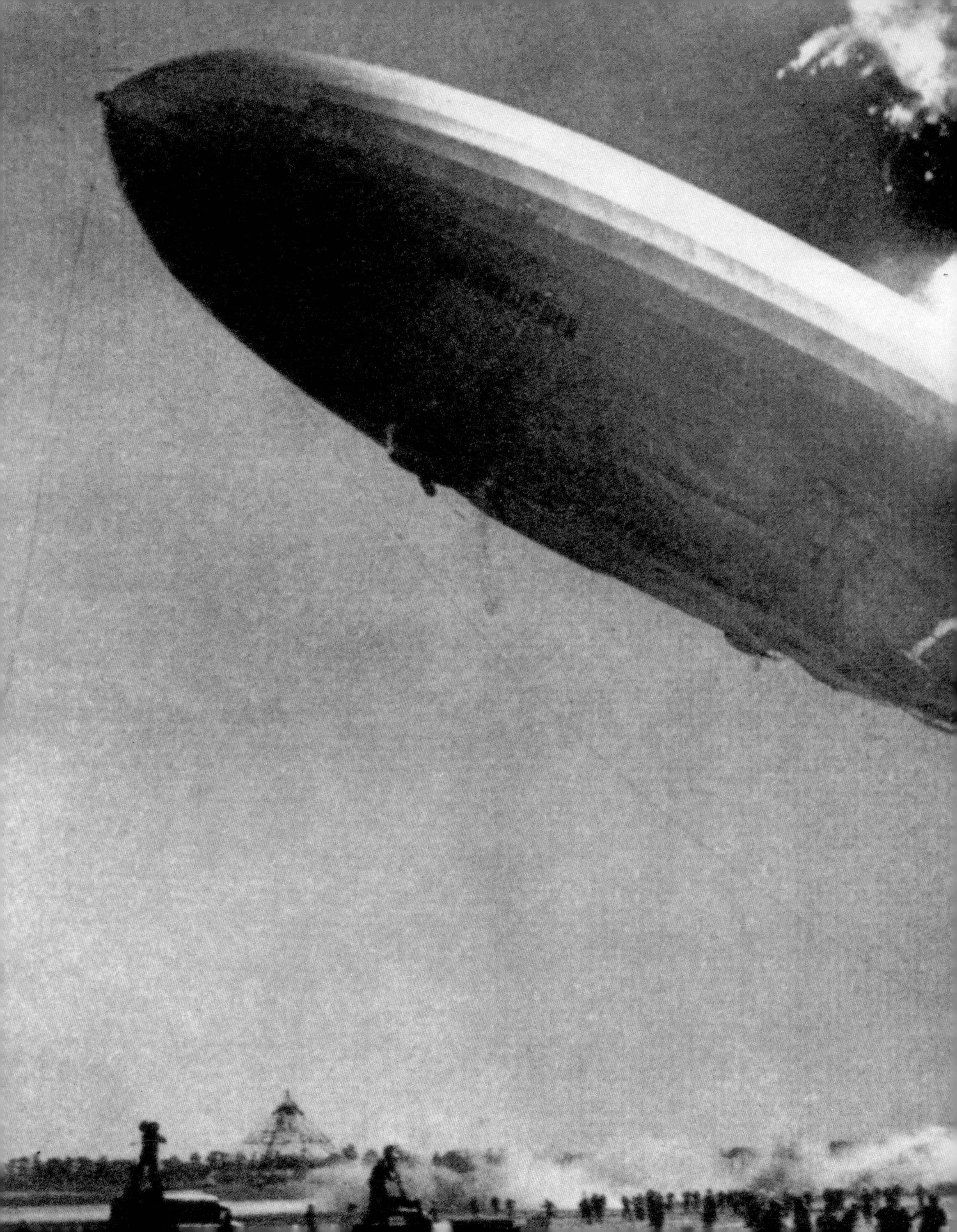

BARELY HALF A MINUTE LATER THE BLAZING AIRSHIP CRASHED TO THE GROUND.

contemplate the notion of sabotage as a cause of the airship's destruction, and the FBI unable to find any evidence of sabotage to support such a theory, another cause had to be found. It was obvious to those on the ground that the *Hindenburg* had not been struck by lightning. Although there had been a storm just before the landing, the weather had been improving and there had been no lightning present.

Having discounted both sabotage and lightning, that really left only one other possibility: a hydrogen explosion caused by a static discharge. But how had the static electricity reached the hydrogen?

During their investigations, the inquiry team learned about the large amount of ballast that had been dropped just before the final landing in an attempt to balance the rear of the airship, which had suddenly become heavier. Eckener suggested in his report that this sudden heaviness might have been due to a loss of lifting gas caused by a rupture in one of the rear gas cells. As a possible cause of the rupture, he pointed to a full speed turn that Captain Pruss had made just before the final approach to the mooring mast. This, claimed Eckener, may have caused one of the bracing wires that held the rigid structure of the airship together to snap, slashing open one of the rear gas cells. This explained where the highly explosive hydrogen gas might have come from, but what about the necessary spark to ignite the gas? Enter Professor Max Dieckmann, expert on all matters electrostatic.

Dieckmann noted that although there was no lightning during the landing attempt, the recent thunderstorm would have created a difference between the electrical charge in the clouds above the airship and the ground below. This potential difference could not have been a problem for the airship, however, unless it had somehow become electrically grounded. Unfortunately, by dropping huge amounts of ballast after the mooring ropes had been lowered to the already wet ground below, Captain Pruss had, in effect, turned the ropes into electrical conductors. With the airship now possessing the same electrical charge as the ground below, conditions were ideal for a discharge of static electricity from the atmosphere surrounding the *Hindenburg*. Under these circumstances, it was only a matter of time before the spark of electricity found the leaking hydrogen gas and the *Hindenburg*, along with the beautiful dream of a sky filled with airships, came to an explosive and deadly end.

... it was only a matter of time before the spark of electricity found the leaking hydrogen gas and the Hindenburg, along with the beautiful dream of a sky filled with airships, came to an explosive and deadly end.

The American investigators largely agreed with the conclusions reached by Eckener and Dieckmann, recording it as the most probable cause of the disaster. To this day, various people continue to put forward all manner of theories as to the "real" cause of the destruction of the *Hindenburg*. In reality, however, the cause is hardly of any consequence. The fact remains that, for whatever reason, too many lives were lost that day and the world would never again look at the airship in the same way.

Chapter 7

The Legacy

A YEAR AFTER THE *HINDENBURG'S* CRASH CROWDS SALUTE THE LIFT-OFF OF THE *LZ130*, THE *GRAF ZEPPELIN II*.

Hermann Goering was not shy when it came to expressing his views on airships: he called them "gasbags". As air minister under Adolf Hitler, he had little choice but to go along with his Führer's almost whimsical interest in the airship.

And much as he despised the weasel-like Joseph Goebbels, there was no denying the propaganda value of the *Hindenburg* and her sister ship the *Graf Zeppelin*. One could be forgiven for thinking that Goering would seize the opportunity that the destruction of the *Hindenburg* presented to do away finally with the airship altogether, but his reaction was quite the opposite.

Through gritted teeth he explained to Hugo Eckener that now, more than ever, Germany must continue with its programme of airship building. The Fatherland could not be seen to end its production of airships as a result of a spectacular failure in a foreign country. It hadn't helped Goering's temper any that the entire world had been able to see the great symbol of the might of the Third Reich crash and burn in such a fashion, the first-ever air disaster caught on camera for a universal cinema audience.

At the time of the crash of the *Hindenburg*, a new German airship was already in production. Called the *LZ130*, it was almost identical in design to the *Hindenburg*. In the wake of the crash at Lakehurst, however, no one was willing to risk another hydrogen-fuelled explosion. With this in mind, it was decided to refit the *LZ130* so that it could be filled with helium. The only problem that remained was how to obtain the helium. At last it seemed that the Nazis had found a use for Hugo Eckener, and the champion of the airship was sent to America to plead the case for exporting helium for non-military purposes. In the meantime the necessary modifications were made to the *LZ130*, which included reducing the cabin capacity to 40 passengers from the original 72. It was clear from this action alone that the *LZ130* was never intended to turn a profit – there simply wasn't going to be enough room for paying passengers on board. But no one was really concerned with profits at this stage, they just wanted a German airship that could dominate the skies once more – only safely this time.

Eckener succeeded in his mission to get the Americans to relax the restrictions imposed by the Helium Control Act, but before the airship was

HAROLD ICKES, US SECRETARY OF THE INTERIOR, PLACED AN EMBARGO ON THE EXPORT OF HELIUM FOR FEAR THAT HITLER WOULD USE IT IN MILITARY AIRSHIPS.

completed in 1938, German troops were marching through the streets of the Austrian capital and the world was left in no doubt as to the expansionist plans of Adolf Hitler. Convinced that the helium intended for the *LZ130* would be used for military purposes, the US Secretary of the Interior, Harold L. Ickes, refused to allow the gas out of Texas. Aware that few passengers would be prepared to travel on a hydrogen-filled airship, the Germans none the less filled the *LZ130* with the inflammable gas and announced that it would be used for training purposes until such time as the Americans could be persuaded to part with their helium.

Non-rigid airships, known as blimps, made a brief comeback in the 1950s as part of the early-warning systems set up during the Cold War.

Renamed the *Graf Zeppelin II*, the *LZ130* was, in reality, destined for a life as a propaganda tool and as a means of spying on the defence capabilities of other European countries. When the Second World War finally began in 1939, however, the Americans had still not handed over the helium that would have made the *Graf Zeppelin II* a viable military aircraft. Rather than send up any hydrogen-filled airships to be shot down in spectacular fashion by the Allies, Hermann Goering eventually had all existing German airships, including both the *Graf Zeppelin I* and the *Graf Zeppelin II*, broken up to be used as raw materials for combat aeroplanes. On May 6, 1940, Goering added gross insult to injury by

blowing up the airship hangars at Frankfurt – three years to the day that the *Hindenburg* had exploded into flames over Lakehurst. (The airship hangars at Friedrichshafen were destroyed by Allied bombing in 1944.) The highly skilled craftsmen who had been responsible for producing the wonderful zeppelins were eventually put to work constructing the dreaded V-2 rockets. Germany, for so long the world leader in airship construction, now abandoned the pursuit of lighter-than-air flight to concentrate on less elegant means of taking to the air. The dream was finally over.

Although the Allies found a limited use for airships during the Second World War, mostly as anti-submarine escorts for transatlantic convoys, the age of the airship as a serious rival to the aeroplane really had come to an end. Shortly after the war, the Goodyear Company enlisted the help of Hugo Eckener in their efforts to revive interest in the airship with the planned construction of a massive, helium-filled craft. At 950 feet long and with a gas capacity of 10 million cubic feet, the proposed airship would have been a truly majestic sight in the sky over New York, but it was not to be. The world simply was not interested in building any more giant airships.

During the 1950s, the blimp – the non-rigid cousin of the rigid airship – made something of a comeback. The Korean War saw the Americans using the blimp for reconnaissance missions, and as the Korean War gave way to the Cold War, blimps fitted with early-warning radar systems were sent aloft in case they were ever needed to detect raids by Soviet nuclear bombers. It was also during this period that the largest non-rigid airship ever built was pressed into service with the US military. The *ZPG-3W* had a gas capacity of 1.5 million cubic feet and was just over 400 feet in length. Although the US Navy built four of these blimps, changes in the way nuclear wars were likely to be fought caused a cancellation of the entire Navy blimp programme. Given that the Soviet Union might be expected to attack mainland America with intercontinental ballistic missiles rather than nuclear bombs carried on board huge bombers, this seemed like a perfectly sensible action. The US Navy's decision more or less brought to an end the military use of the airship anywhere in the world.

As the 1960s got under way, the number of airships travelling through the world's skies continued to decline to an alarming degree. By 1962, it is generally believed that there were perhaps only two airships still in operation, and they were both blimps.

The economic crisis of the 1970s came as a surprise to many people who had believed that the world's resources were inexhaustible. They had spent several decades filling their highly inefficient cars with cheap fuel, safe in the knowledge that there was always plenty more to be had at the filling station. As the oil producers began to put the squeeze on the rest of the world, cheaper, less oil-dependent means of travel were once again investigated. Aeroplanes may be powerful and fast, but they use more fuel during the first hour of flight than an airship would use to travel several thousand miles. With this in mind, several government bodies around the world began handing out grants for the development of cheaper, alternative means of travel. Soon brand-new airship designs were being touted around, including one proposal for an airship so large that it would have been necessary to build it in a converted canyon – covered over with a temporary roof – as the cost of building a hangar for the craft would have made the project prohibitively expensive.

In a more realistic and certainly more interesting development, some engineers came up with hybrid designs. These were part airship and part aeroplane or, in the case of one design that actually got off the ground, part airship and part helicopter.

Frank Piasecki, the designer of the part airship/part helicopter option – which he called the Heli-Stat – hoped to combine the lift capacity of an airship with the manoeuvrability of a helicopter. To this end he strapped four

THIS LETHAL ROCKET WAS ONE OF MANY MADE BY THE FORMER ZEPPELIN WORKERS.

army surplus helicopters to an old US Navy blimp. The idea seemed sound enough, but during a test flight in July 1986 at Lakehurst, site of the *Hindenburg* crash, pilot error resulted in four injuries and one death. The Heli-Stat was never flown again.

Several more hybrid designs were attempted but most either never left the drawing board or failed to get off the ground. Some, such as the frankly bizarre-looking Cyclo-Crane, even suffered the indignity of ending up as curios in museum collections. Since the 1980s, however, there has been a resurgence of interest in the airship, although many of the designs have tended towards the conventional and usually resemble, at least in shape, the blimps that were around 60 years ago.

For the most part, these new blimps have been used for advertising purposes, although some have been employed for aerial camera work, covering events such as football games and large parades. A few are even available for hire for sightseeing trips.

THE AERO-STAT BLIMP WAS A HYBRID DESIGN CONSISTING OF PART AIRSHIP AND PART AEROPLANE.

The new millennium, however, has proved to be a spur to the imaginations of designers and dreamers the world over. For the first time in over 60 years, people are talking about a new age of the airship – and for once, it is not all talk. In ten countries around the world, fifteen different companies are engaged in attempts to restore the mighty airship to the skies.

Leading the way, with the production of a massive new craft, is South Africa's Hamilton Airship Company. Their new ship, named the *Nelson* in honour of the country's political and spiritual leader, is an impressive 460 feet in length and is designed with the intention of recreating the splendour of the great days of the zeppelin. Passengers will be accommodated in luxurious suites, but also have the option of wandering around the airship's three observation decks, with their restaurants and cocktail lounges. The stated aim of the airship manufacturer is to produce a craft that will introduce a new era in leisurely and safe air travel.

Cheaper to operate than a Boeing 747, the *Nelson* has been designed to withstand wind speeds of 100 miles per hour. As a further safety feature, most of the ship has been constructed from a composite of carbon fibre and kevlar, which is used to contain the helium lifting gas. Because of its unique docking system, the *Nelson* can be landed on a space no larger than a heliport. With a range of 6,000 miles, and the ability to stay airborne for 112 hours, the *Nelson* is intended as a passenger cruise ship. The military, however, are looking once again at using the enormous lifting capacity of the airship to transport troops and equipment to the battle zone.

Over six years in development by Lockheed Martin – a company normally associated with Stealth aircraft – the Aerocraft is the most ambitious airship/ aeroplane hybrid ever attempted. Designed to allow the rapid deployment of large quantities of heavy military equipment, the Aerocraft is a

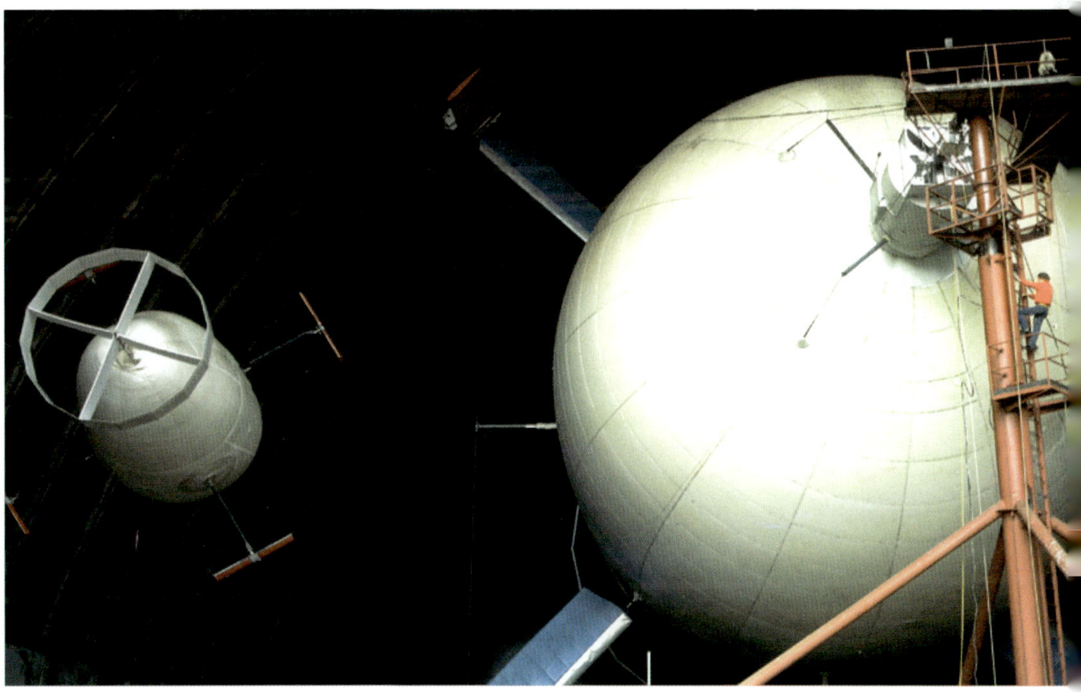

THE CYCLO-CRANE – AN INGENIOUS IDEA THAT NEVER REALLY TOOK OFF.

superblimp in every sense. At nearly 800 feet long and 250 feet wide, the new craft has the carrying capacity of 14 Boeing 747 jets. It achieves this by combining both old and new technologies.

The upper part of the Aerocraft has a lightweight skin containing thousands of small pockets of helium, just like a conventional airship. The lower part of the craft, however, resembles an ordinary, if rather large, aeroplane. The hybrid is powered by four "tiltrotor" engines which, as the name implies, are designed to provide both forward thrust and lift.

TODAY AIRSHIPS ARE USED MAINLY FOR ADVERTISING PURPOSES; THIS ONE DRIFTS GRACEFULLY OVER THE ROSEBOWL STADIUM IN PASADENA, CALIFORNIA.

Capable of carrying 500 tons of military hardware – principally tanks and heavy artillery – the Aerocraft would be flown to the edge of a war zone and its cargo unloaded, ready to go into battle. Fears that the airship would be especially vulnerable to attack were dispelled in a memorably brief, but very convincing demonstration of its safety, when a soldier from the British Army fired forty shots into the craft without causing any serious damage.

Also under consideration by the military is a new airship being built in Britain by the Airship Technologies Group. When completed, the *AT-04* will be the largest airship to be built since the *Hindenburg* fell from grace in 1937. Although designed primarily with the military in mind, the *AT-04* can be adapted for civilian passenger services.

Built from modern, lightweight, extremely strong materials, the *AT-04* will be seven storeys high, 270 feet long and have a helium gas capacity of 500,000 cubic feet. Powered by three 350 horsepower engines, driving reversible-pitch propellers, the craft will cruise at around 50 miles per hour, but will have a top speed of 80 miles per hour. One of the engines will be fitted to the stern of the ship, while the other two will be mounted on each side of a 60-foot gondola. Each of the engines can be rotated through 90 degrees to aid manoeuvrability at low speeds.

Uniquely, the airship is to be fitted with a special "bow thruster". This device, when combined with the effects of the movable engines, will give the crew a degree of control over the airship that would have been beyond the comprehension of early zeppelin pioneers. Because of this, it will be possible to land the *AT-04* without the help of a ground crew, although usually there will be at least one handler standing by in case of emergencies.

When cruising, the *AT-04* will use just the rear engine. This offers the advantages of conserving fuel while at the same time dramatically reducing any unwanted noise or vibration. Intended for service with the US Navy, the *AT-04* will most likely serve as an airborne radar platform. At present, airborne radar is carried by aeroplanes capable of staying in the air for a few hours. By contrast, the *AT-04* will be able to remain aloft for days, or even weeks, at a time.

The Airship Technologies Group is also looking at ways of using the *AT-04* for civilian commercial purposes, and to this end will be offering the ship to various organizations as a huge airborne television screen.

So it would seem that, despite earlier dire warnings, the airship is far from dead. There is a genuine desire among many people for a return to the days of leisurely and stately air travel, and the

WHEN COMPLETED THE *AT-04* WILL BE LARGER THAN THE *HINDENBURG* AND ITS MOVEABLE ENGINES WILL ENABLE IT TO LAND WITHOUT THE USE OF A GROUND CREW.

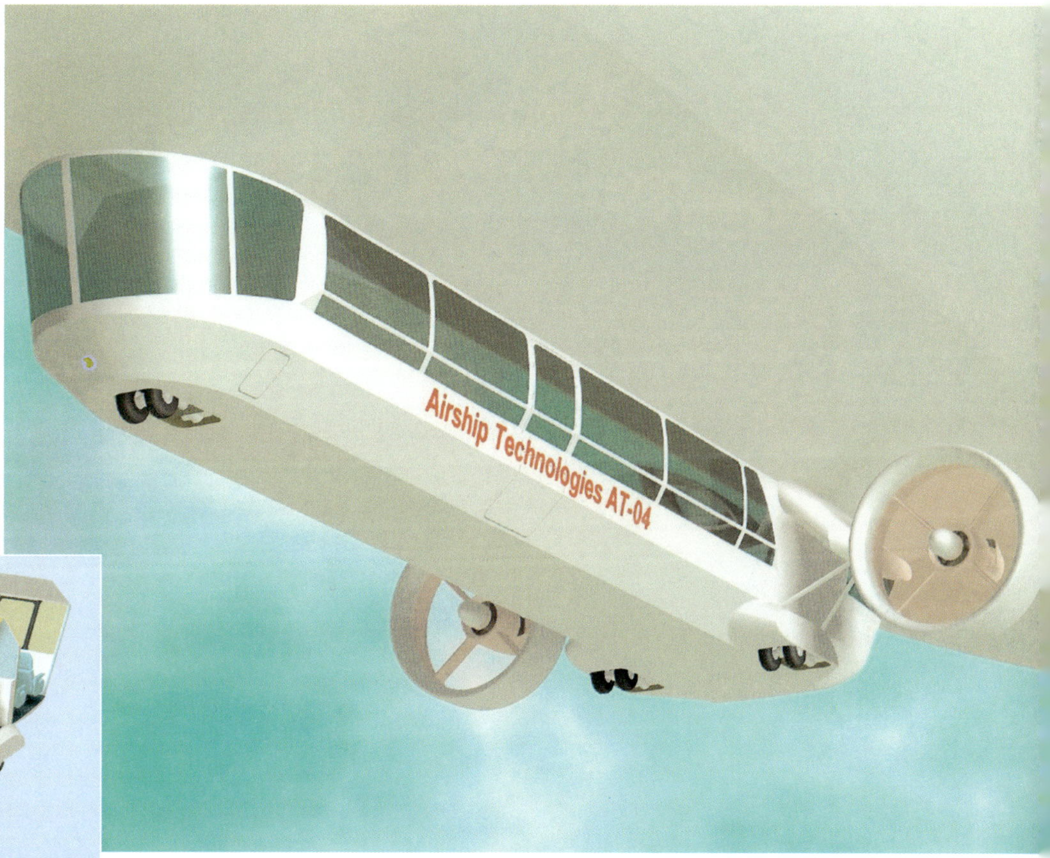

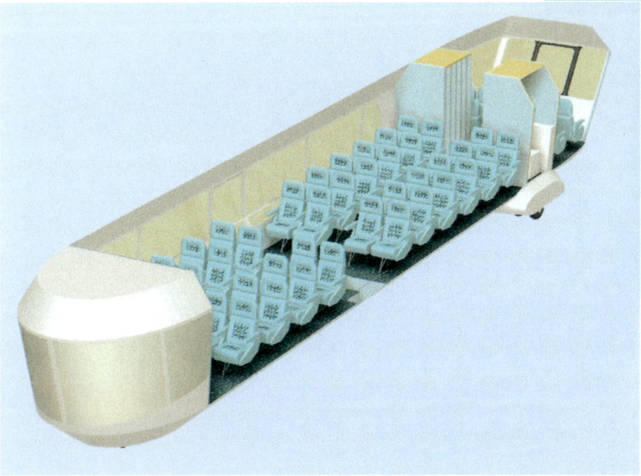

A BLIMP GLIDES THROUGH THE SKIES OVER NEW YORK.

airship is the obvious means of achieving this. While there will always be a demand for ever more rapid methods of travel between continents, the attractions of the airship for those with a little more time on their hands are obvious. After all, only the dimmest of imaginations would reject the opportunity to cruise slowly over the Pyramids, follow the migration of wild animals across the Serengetti, or see the Alps as only an eagle can.

The unfortunate decline of the airship in the middle years of the twentieth century represented a victory of economics over beauty, of hard cash over sweet dreams. Had they lived to see it, men such as the Montgolfier brothers, Henri Giffard, Count Ferdinand von Zeppelin and even Hugo Eckener, who invested their entire lives in the construction of these marvellous machines, would have thought that the human race had lost its ability to dream. But, with the fortunes of the airship once more in the ascendant, it is clear that this is not the case.

The airship is the embodiment of grace and beauty. Its ability to fire the imaginations of all who see it ensures that it will never be lost to us. As long as we maintain a sense of wonder and of pride in the achievements of the human race, there will always be a place for the airship. Anyone who has ever gazed in awe at the magnificence of these craft knows this to be true. In centuries to come, our descendants will look back at the airship and realize that here was a people who could dream and had the courage to make their dreams reality.

THE *AT-10*, FITTED WITH A HUGE T.V. SCREEN, SAILS OVER THE RIVER THAMES IN LONDON.

Index

Picture Credits

The publishers would like to thank the following sources for their kind permission to reproduce the pictures in this book:

AKG London 3, 7, 9b, 12, 18, 24, 26t, 27t, b, 28-9, 35, 45, 50, 56, 57, 58, 60, 62, 64t, b, 65, 66, 82, 84
Courtesy Airship Technologies 93, 95
Bridgeman Art Library/Bonhams London, UK *The Fall of Icarus, Jacob Grimmer (c.1526-89)* 30
Jean-Loup Charmet 10, 11, 13, 17
Corbis 6, 31, 91/Bettmann 14, 15, 20, 21, 26-7, 28, 85/UPI 25t, 32, 33, 39, 41, 42, 44, 46, 47, 68-9, 73, 86; Owen Franken 94; George Hall 90; Dave G Houser 94; National Archives 51, 61, 67, 86-7; Joseph Sohm/Chromosol 96; US Army White Sands 89
Hulton Getty 23, 25b, 34, 36, 37, 38, 48, 53, 55t, b, 71, 78
Image Select/Ann Ronan 8, 9t, 16t, b, 22
Tony Stone Images/Ken Biggs 92
Topham Picturepoint 49, 52, 54, 59, 74-5, 75b, 76, 79, 80-1

Every effort has been made to acknowledge correctly and contact the source and/copyright holder of each picture, and Carlton Books Limited apologises for any unintentional errors or omissions which will be corrected in future editions of this book.